To our friend Emanuele Rimini

"From error to error one discovers the entire truth"

S. Freud

Preface

The book contains selected contributions, extracted from the Theses produced by the people that attended the first edition of the Master in Microelectronics and Systems that has been held at the University of Catania from March 15 1999 to March 15 2000.

The mentioned Master has been in particular organized among the various activities of the " Istituto Superiore di Catania per la Formazione di Eccellenza ". The first aim of the Master in Microelectronics and Systems has been to have an advanced course in the specified topics for a selected numbers of students with the Laurea Degree in Physics or Electrical Engineering with the scope to give at the attending people a complete view in the emerging areas of Microelectronics like as CAD, Microprocessor Design, Circuits for Wireless Communications, VLSI Circuits for Intelligent Systems, Electronics for Automotive Systems. The second aim has been to involve in the activity companies and research government groups, working in microelectronics projects to contribute to the Master activity establishing a real world link with advanced problems in the related area. In fact the Master included both a period of intensive lectures and a period of stages in companies or in laboratories where they established the key points of their final thesis.

The project has been founded by MURST and organized by the University of Catania in collaboration with STMicroelectronics that also supported the initiative with other companies like ACCENT and ITEL. To summarize, the aim of the School has been twofold:

- ✓ to provide students with Laurea to furtherly improve their skills in advanced topics of Microelectronics and System's design;
- ✓ to bring together the experience from academia and industry cooperating for an high level educational project devoted to reach high professional levels.

Besides the tutorial activity during the teaching hours, provided by national and international researchers, a significant part of the School has been dedicated to the presentation of specific CAD tools and experiments to directly prepare the students to solve specific problems during the stage period and in the thesis work.

We would like to remark that the stage themes and the thesis subjects have been accurately chosen in accordance with some specific emerging topics and we would like to thanks both the tutors of the University and those of companies that encouraged and stimulated the activity of the students that reached high level of preparation and presented in their theses often innovative solutions.The main results are reported in this book.

The material collected in the volume can constitute the basis on some key subjects of interest both in the University courses for undergraduate or graduate and for researchers and designers in companies.

The book will be diffused in the R&D Divisions of the companies involved in the initiative in order to give the status of some research projects and to the next persons attending the 2nd Edition of the Advanced School that is just starting now.

To conclude we would like to thank all the people that gave their foundamental contribution for the success of the initiative. A particular thanks will be for Dr. Michelangelo Pisasale and for Dr. Bruno Ando' for their appassionate help that we received in preparing this volume.

Prof. G. Ferla
DSG R & D Director, STMicroelectronics S.r.l. - Italy

Prof. L. Fortuna
Istituto Superiore di Catania per la Formazione di Eccellenza, Universita' degli Studi di Catania - Italy

Dr. A. Imbruglia
DSG R & D Rad-Hard Product Development Manager, STMicroelectronics S.r.l. - Italy

Contents

Analog Layout Area Optimization

Giuseppe Francesco Giuffrida

DEES, Engineering Faculty, University of Catania

Viale A. Doria 6, 95125 Catania - Italy

Introduction

Chip-Area optimization is one of the most important objective for silicon foundry. In particular, it looks like a very hard task for analog device implementation. Chip-Area optimization goal involves a large number of topics and problems, such as performance degradation, matching and symmetric constraints, parasitic extraction.

The layout of analog circuits is intrinsically more difficult to be obtained than the digital one. Besides, analog circuit performances are much more sensitive to physical implementation details than the digital ones [1]. Custom design requires great flexibility, and layout synthesis is often a multiple-objective optimization problem. Along with area, wiring length and delay, two groups of relevant issues must be taken into account: topological constraints (i.e. symmetries and matching) and parasitics (associated with devices and interconnections).

Unconstrained compaction can degrade parasitics by modifying the spacing between interconnections and matched devices. Both asymmetries and device mismatches along with the interconnections can introduce intolerable performance degradation [2]. Since these effects are unavoidable, the main task in Computer-Aided Analog Layout Synthesis is to control parasistics on circuit performance and to keep the layout-induced performance degradation within user-defined margins.

On the basis of the above consideration, suitable area optimization algorithms are very hard to be implemented. Moreover, generating a high-performance analog layout results in a both difficult and time-consuming task. Several approaches to automate analog layout have been developed. In most of them typical methodologies derived from digital world have been used. Generally, these approaches have not achieved the same layout density and flexibility obtained by expert designers.

Other procedures are characterized by a knowledge-based approach, or required the user to provide basic hints on the layout structure.

Physical design implementation without explicit reference to circuit performance specifications is the main limitation of the above approaches. When the layout is designed, the main task consists of extracting the layout parasitics and simulating the circuit to check if the performance specifications are met. Anyway, no information about the causes of the eventual failure is systematically available. Usually, it results in a large number of expensive layout extraction-simulation iterations.

Moreover, a great importance to parasitic control during routing is given; capacitance bounds between nets are preserved by setting the minimum separation between wire segments in a channel, and by ordering them to avoid crossover (if possible!). In particular, cross-coupling minimization is a routing target, while stray resistances are usually controlled by means of variable wire segments widths.

In [3] very effective analog-specific placement and routing tools is provided, but no explicit reference to the performance specifications is given. Moreover, area routing and unconstrained placement with abutment capability (based on weighted parasitic minimization and matching constraint enforcement) provide the layout with high flexibility and good area performance. However, these tools give no instructions on both parasitic weights definitions and matching constraints (these results must always be supplied according to users' experience and knowledge on circuit behavior).

Among bibliographic references (IEEE libraries) [4] [5] [6] [7] two approaches seem to be the best allowing area optimization procedures: the performance-driven approach [1] [2] (based on a Simulated Annealing procedure for placement) developed by Prof. A. Sangiovanni-Vincentelli's team, and a Hybrid Genetic Algorithm for Constrained Placement Problems [8] [9] proposed by Schnecke and Vornberger.

Experimental Activity

Introduction

Both the approaches investigated in the previous section have been taken into consideration. In order to propose some area optimization procedures, a careful description of either the performance-driven methodology or the genetic algorithm application was given to the RF and R&D teams. Anyway, the search for an automatic placement licensed tool available in ST Microelectronics was encouraged.

As a consequence, the latter part of the stage activity was dedicated to the investigation of licensed software in ST [10] [11]. In particular, a feasibility study of a suitable tool was carried out and both the elaboration and coding of an algorithm for layout area measurements was carefully performed.

Automatic placement tools in ST Microelectronics

In order to properly fix an automatic placement tool, licensed software in ST was examined. Due to the performance to be assured by the required software, DLP (*Device Level Placer*) in the *Cadence release 4.4.2* of the suite *Design Framework II* has been taken into account. It should be underlined that new versions of Cadence suite is out of this kind of tool.

Of course, the necessity of a feasibility study for the application to the analog project flow arises. As benchmarks for the feasibility study, ST devices developed in HSB2 technology for Radio Frequency applications and CMOS devices (i.e. operational amplifiers) implemented by the author were used.

At this point, a brief description of the software is necessary.

The *Device-Level Placer* (DLP) should automatically place devices. The placer lets the designer add constraints to the devices before placement and then should use the constrains while placing the devices. The placer is made up of two different and independent tools.

The first tool is a partitioner-based placement tool activated by selecting the *Place Using Constraint-Driven Placer* option, as illustrated in Figure 1. The partitioner models each component as a dot with all pins in the center. It places the components in the design area while trying to minimize the net crossing between geometrically partitioned areas of the design. This tool is deterministic: it always produces the same solution. The partitioner should always follow constraints if possible.

The second tool is a localized low-temperature simulated annealing optimizer activated by selecting the *Optimize* option, as illustrated in Figure 1. If this tool has not reached an optimal solution in the specified time, it uses the best placement it found during that time. The optimizer is non deterministic: the solution it produces varies from run to run. The Automatic Placement and the Constraint Editor form are shown in Fig.1. Manageable constraints are: Distance constraints, Alignment constraints, Grouping constraints, Symmetric constraints, Fixed constraints.

When DLP has finished, the Automatic Placement form reports overlaps, devices remaining outside the design boundary (*prBoundary*) of the layout window, and constraints still unsatisfied. If DLP is unable to satisfy constraints, it should print one or more messages to the text window in the Automatic Placement form listing the devices outside the *prBoundary*, overlaps, and constraints not satisfied.

It should be observed that the required constraints could not be satisfied due to either large time-consuming or overconstraints effects (set of constraints that cannot be simultaneously met). By using the *Constraint Editor*, it is possible to set a number of constraints to be releasable. By giving enough time (how much?), the placer should place all components satisfactorily and begin to work on the overconstraints. To solve this problem, DLP should relax the releasable constraints affecting the greatest number of objects. If suitable solution is not still obtained, relax of unreleasable constraints may occur.

The Placer section of DLP seems to be quite satisfactory. It gets no long time to fix the placement in the respect of the required constraints. Many successfully outcomes have been obtained with the above-described benchmarks. But, very often the manual placement get less time then the automatic approach, due to the required time for constraints setting. This states as an important drawback for the tool. Probably, a user-friendly interface could allow this software to be much more acceptable for the analog project flow.

On the other hand, the optimizer section cannot be used because of significant output are obtained in a large amount of time, independently on the circuit complexity. As an example, to obtain a good automatic placement of a 20-transistors device more than two hours are required. Moreover, if a small time small optimization time is set, fundamental constraints relax unavoidably occur.

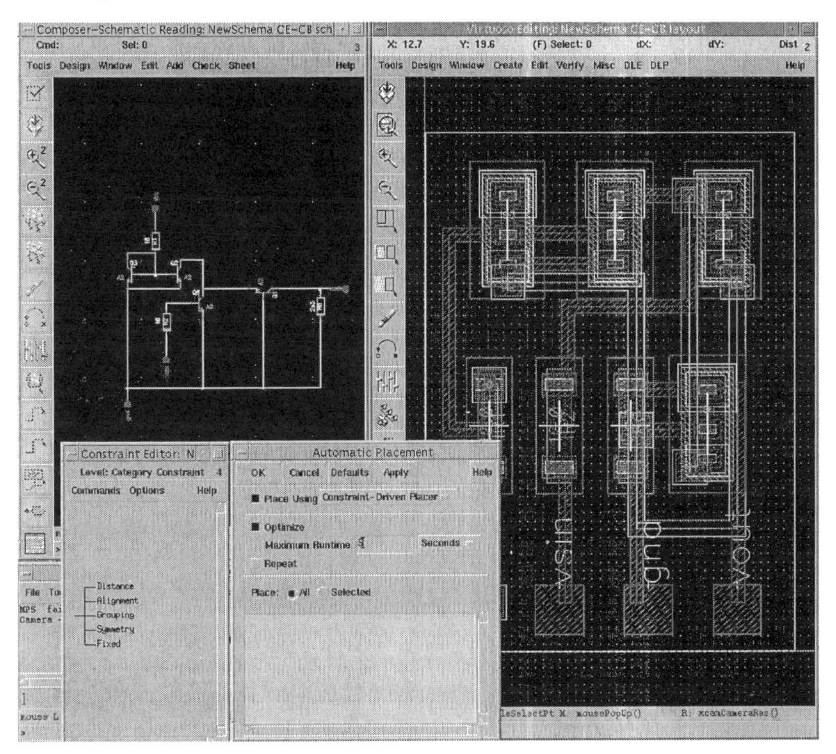

Figure 1: The device-level placer

For the above reasons the feasibility study produced negative outcomes. Leading so to the consideration that DLP tool cannot be integrated in the analog project flow. As a matter of fact, it has never been used by the ST project team and it is not implemented (that is a very significant fact!) in the current version of *Cadence* suite.

Software for areas evaluation

The conclusive section of the stage period wad dedicated to the elaboration of a procedure for occupied areas evaluation. This procedure may be considered as the first step for the definition of a layout compaction methodology. In fact, instead of an *a priori* approach (automatic placement), an *a posteriori* one was proposed.

Since the research for an automatic placement tool among all licensed software in ST provided no interesting results, the use of iterative compaction is necessary. However, good compaction can be obtained only if an exact occupied areas evaluation is available. Actually, this approach was suggested by Eng. Giuseppe Ferla (R&D Chief) in order to obtain equal partition of the occupied areas into fundamental layers. This goal could ensure a better utilization of methodologies employed in the chip generation process.

To fit the target another investigation on ST licensed software was necessary.

In order to select an areas evaluation tool, a deep research activity among all licensed software in ST was developed. As a result, only one tool was available: *Design Statistics* in DLE (*Device Level Editor*). It belongs to the *Cadence release 4.3.4* of the suite *Design Framework II*.

Unfortunately, this tool is inadequate for the considered purposes. Since it returns the instance cells areas, it furnishes no reliable output. In order to simplify manual placement, the instance cells are realized with a boundary layer, which extends the real cells dimensions about an half of the minimum distance between cells (in HSB2 technology this enlargement is nearly 3μm). Since the boundary layer generally results concentric to the cells, it generates a large outlined area. Hence, *Device Statistics* carries out redundant evaluations. As an example, for no dense layout it returns occupation index greater then 100% (that's practically unacceptable!).

Figure 2 shows some instance cells, where the boundary layers are represented by means of white rectangles. Obviously, these cannot be used for active areas evaluation.

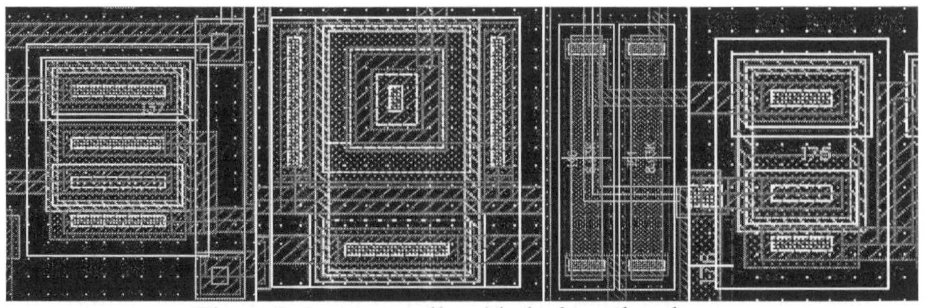

Figure 2: Instance cells with the boundary layer

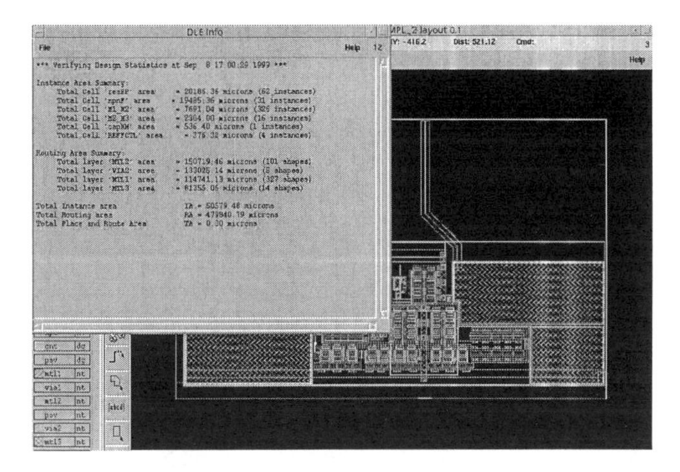

Figure 3: The design statistics in DLE

Besides, Device Statistics gives no reliable results if a specific layer (*prBoundary*) is not previously defined.

In Figure 3 the form of Design Statistics is reported. It is divided into two different sections: Instance area summary, Routing area summary. At the bottom, a comprehensive synthesis is displayed. The layout in Figure 3 has a 0 μm value for the Total Place and Route Area. This absurd result comes out from the absence of the *prBoundary* layer.

Figure 4 shows Design Statistics after using the *prBoundary* for the same layout. For the Total Place and Total Route Area a value greater than 0 together with an Utilization index are displayed. Are they acceptable values? May the reported layout have an utilization nearly to 98%? Is there only a 2% waste area?

Due to the use of boundary layer cells, parts of the routing areas are twice computed in the utilization index.(see Figure 2). In fact, all the routed metals inside this layer are at the same time both active and routing area.

On the basis above consideration, a new procedure has been implemented.

Areas evaluation software: the new developed procedure

A novel procedure for areas evaluation was coded in SKILL (inside CAD suite *Cadence* language). It consists of three different parts.

The first one concerns the user interface, which is really user-friendly: some dialogue windows drive the operator during the whole activity.

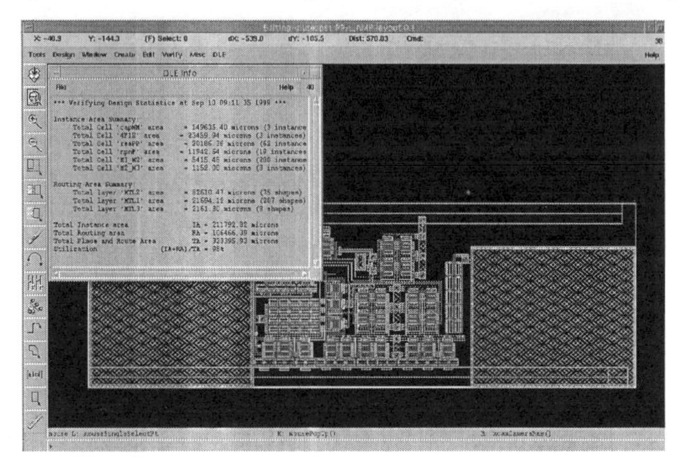

Figure 4: Design Statistics after using prBoundary

A specific dialogue form, reported in Figure 5, asks to choose between restricted investigation areas or the whole layout option.

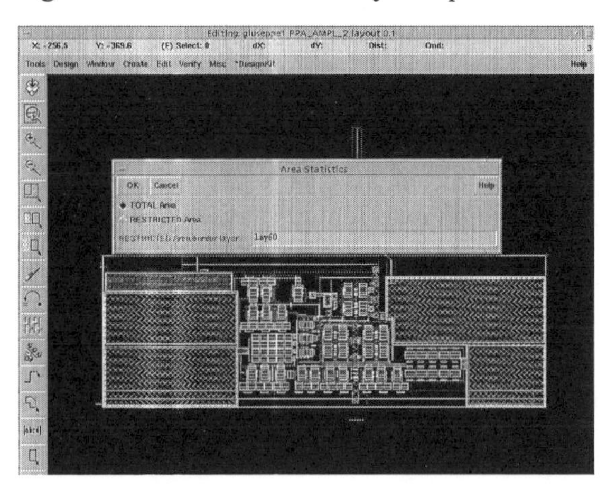

Figure 5: A dialogue windows in the new procedure.

In the first case, it is necessary to bound the desired areas with a specified layer. As an example, in Figure 5 the *layer60* (one of the free-use layers) has been chosen.

Combinations of multiple selected areas are allowed. In this case a window informs the users about presence of more than one shape in the specified layer and asks for continuing calculation. Choosing the yes option, the evaluation will set the OR among all the selected areas.

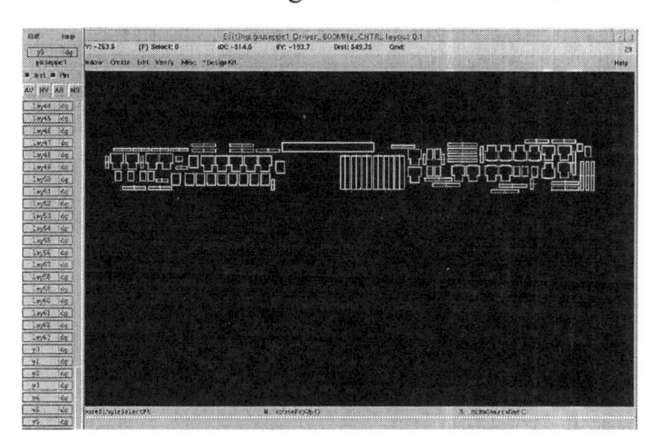

Figure 6: Active area saved in layery9

The second section is related to the interface between SKILL and DIVA. This last is an inside *Cadence* suite robust tool devoted to layers extraction from a given layout. For this tool a graphic interface is not supplied. By combining layers of the occupied areas, it is possible to extract the boundaries of active areas, capacitances, routing, and pads. Obtained boundaries are saved into particular free-use layers and then insulated from the remaining layout. As an example, in Figure 6 the active area saved in the *layery9*, is shown.

In the third section the areas of extracted layers are evaluated. The tool output is reported in a text window, which can be saved in a file. It reports the investigated areas both in absolute value and in percentage of the selected area. Moreover, area routing is fractionated in each metal. Another important tool output is the utilization index that is given as follows

$$utilization = \frac{active\ area + capacity\ area + pad\ area + routing\ area}{total\ area}$$

Figure 7 reports the procedure output. Practically, the output is a form divided in three parts. At the top, total, active, capacity, pads and routing areas values (percentage and absolute) are reported. In the middle part, the utilization index is shown. At the bottom, routing area is fractionated in each metal.

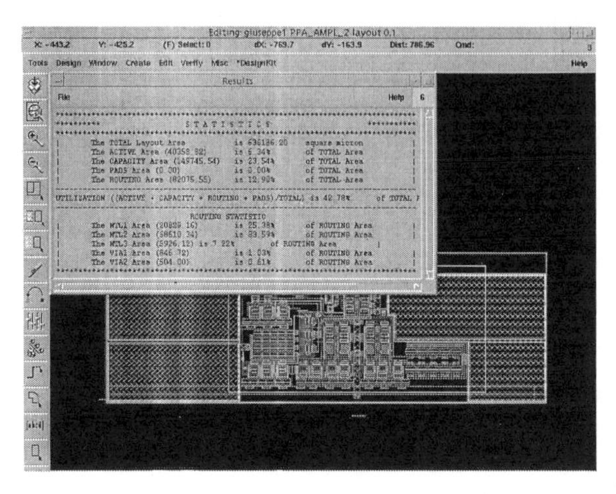

Figure 7: New procedure output

Areas evaluation software: procedure testing

In order to verify its functionality, the proposed procedure has been tested. As benchmarks, devices developed by ST for RF applications have been used. Testing outputs are more than satisfactory: a 6000-transistor device takes, for example, only about 20 seconds for a complete area investigation. Moreover, by using a manual fracturing approach, procedure outputs have been systematically verified.

A comparison between Design Statistics and the proposed procedure is necessary. As shown in Figure 8, differences between Design Statistics and the developed procedure outcomes are relevant.

Areas evaluation software: some details

Since the section devoted to DIVA interface is the most important one, some details about it are reported. The developed procedure generates two DRC files (DRC is the extension of DIVA files), named *Area1.drc* and *Area.drc* respectively. The first one is related to both the extraction and saving (*layery0*) of SUBSTRATE (it looks like an area variable), which is given by either the whole layout area or the selected area. In the first case the instruction *geomBkgnd* is used, alternatively the OR among all parts in the selected layer is computed.

Area.drc file is devoted to the extraction of active, capacity, pads and routing areas. Extracted areas are saved in *layery1-y9*. Obviously, before areas extraction, these layers have to be erased.

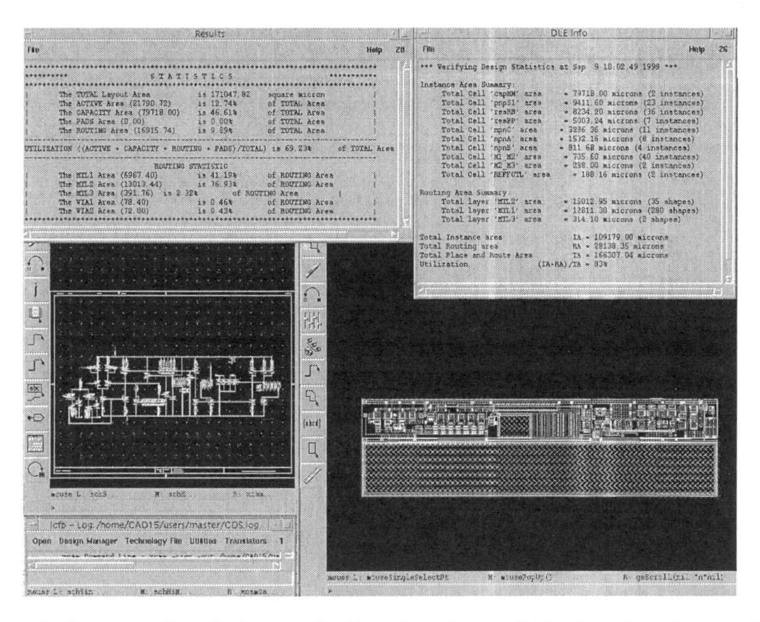

Figure 8: A comparison between Design Statistics and the developed procedure.

Active area concerns with transistors and resistances areas (a successive partition in two different computations is encouraged). It is obtained by the following steps:

UNIDIFF = OR among all diffusions,

UNIDEP = OR among all depositions,

UNION = OR among depositions and the common part between unidiff and unidep,

ISOSI = ISO butting or overlapping with UNION,

CELLE = OR between UNION and filled ISO,

then CELLE is saved in *layery9*. If restricted area option is selected an AND between CELLE and the selected layer is computed, before saving.

Pads area is extracted by the following steps:

MTLPSV = all PSV enclosed in MTL1,

PADS = 15μm^2 enlargement (due to ISO and MTL1) of MTLPSV,

then PADS is saved in *layery8*. If restricted area option is selected an AND between PADS and the selected layer is computed, before saving.

Capacity area is determined by the following steps:

capmetal = all metalcap enclosed in mtl3,

CAPMETAL = 3μm^2 enlargement of,

CAPMETAL = previous CAPMETAL minus AND between previous capmetal and pads,

then CAPMETAL is saved in *layery7*. If restricted area option is selected an AND between CAPMETAL and the selected layer is computed, before saving.

Routing Metal1 area is obtained by the following expression FIRSTMETAL = MTL1 non overlapping CELLE, CAPMETAL and PADS,

then FIRSTMETAL is saved in *layery1*. If restricted area option is selected an AND between FIRSTMETAL and the selected layer is computed, before saving.

Routing Metal2, Metal3,Via1 and Via2 area are similarly obtained and saved in *layery2*, *layery3*, *layery4*, *layery5* respectively.

Moreover, **Total Routing area** is given by an OR among all metals and vias. It is saved in *layery6.*

Obviously, after the writing phase both the files have to be compiled. Successively, they are executed due to the instructions

```
tcLoadTechFile(CurrLib "~/Area1.drc")
ivDRC(?cell getWindowRep() ?full t ?hier nil)
tcLoadTechFile(CurrLib "~/Area.drc")
ivDRC(?cell getWindowRep() ?full t ?hier nil)
```

A specific procedure (CAT_Area) is called to evaluate *layery1-y9* areas.

In order to erase *layery1-y9* from current layout, two new files, containing erasing instructions, are overwrote on previous Area.drc and Area1.drc. These new files are compiled, executed, and finally removed with a *csh* call.

Conclusions

After a section dedicated to the feasibility study of DLP, a new procedure for areas evaluation has been presented. Since the proposed procedure is oriented to Radio Frequency applications, it is devoted to HSB2 technology.

By adapting the DIVA interface section (second section of the developed procedure), generalization to other technologies can be simply obtained.

Evaluation of inductances areas has to be available. Taking into account the RF orientation of the HSB2 technology, this states as a very important goal.

The utilization index may be used to select the best compacted view of the same layout, leading so the proposed automatic procedure. It should run some compaction sequences for a given analog layout. Then, the obtained compacted views can be sorted by means of their utilization index. Obviously, best compactions are allowed for high index values.

References

[1] Malavasi, E., Charbon, E., Felt, E., Sangiovanni-Vincentelli, A., *Automation of IC with Analog Constraints*. IEEE Trans. Computer-Aided Design, vol. 15, no. 8, pp. 923-942, Aug. 1996.

[2] Choudhury, U., Sangiovanni-Vincentelli, A., *Automatic Generation of Parasitic Constraints for Performance-Constrained Physical Design of Analog Circuits*. IEEE Trans. Computer-Aided Design, vol. 12, no. 2, pp. 208-224, Feb. 1993.

[3] Chon, J. M., Garrod, D. J., Rutenbar, R. A., Carley, L. R., *KOAN/ANAGRAM II: New Tools for Device-Level Analog Placement and Routing*. IEEE J. Solid-State Circuits, vol. 26, no. 3, pp. 330-342, Mar. 1991.

[4] Mathias, H., Berger-Toussan, J., Jacquemond, G., Gaffiot, F., Le Helley, M., *FLAG: A Flexible Layout Generator for Analog MOS Transistors*. IEEE J. Solid-State Circuits, vol. 33, no. 6, pp. 896-903, June 1998.

[5] Zu Bexten, V. M., Moraga, C., Klinke, R., Brockherde, W., Hess, K., *ALYSIN: Flexible Rule-Based Layout Synthesis for Analog IC's*. IEEE J. Solid-State Circuits, vol. 28, no. 3, pp. 261-268, Mar. 1998.

[6] Lampaert, K., Gielen, G.,. Sansen, W. M, *A performance-Driven Placement Tool for Analog Integrated Circuits*. IEEE J. Solid-State Circuits, vol. 30, no. 7, pp. 773-780, July 1995.

[7] Säckinger, E., Fornera, L., *On the Placement of Critical Devices in Analog Integrated Circuits*. IEEE Trans. Circuits Syst., vol. 37, no. 8, pp. 1052-1057, Aug. 1990.

[8] Schnecke, V., Vornberger, O., *Hybrid Genetic Algorithms for Constrained Placement Problems*. IEEE Trans. Evolutionary Computation, vol. 1, no. 4, pp. 266-277, Nov. 1997.

[9] Lienig, J., *A Parallel Genetic Algorithm for Performance-Driven VLSI Routing*. vol. 1, no. 1, pp. 29-39, Apr. 1997.

[10] *Device-Level Editor (DLE), Router (DLR), and Placer (DLP)" Cadence Design Framework II handbook*. Nov. 1997, Chapter 7, pp. 1-96.

[11] *Virtuoso Compactor Reference Manual" Cadence Design Framework II handbook*. Nov. 1994, Chapters 1-5.

Analysis of a Flash Memory Device

M. Pisasale[1], S. Poli[1], S. Conte[2]

[1]*STMicroelectronics S.r.l.*

Stradale Primosole 50, 95100 Catania - Italy

E-mail: (michelangelo.pisasale, salvatore.poli)@st.com

[2]*Accent S.r.l.*

Via F. Gorgone 6, 95030 Catania - Italy

E-mail: silvano.conte@accent.it

Introduction

The world of integrated memories is a market in great expansion both for quantities of production and technological evolution: in Computer Science for instance, we find a memory in PC, in printers, in Disk Driver, in Work Stations, where the need of storing and processing information requires high performances both in speed and density. An other immediate application is in telecommunication where the data-transmission through non conventional means (such as mobile networks, GSM) but also through the telephone or internet, in which fax and modem work as interface between the telephone or the Personal Computer and the NET, requires the exchange and thus the storage of information of every kind (from the video to the audio ones).

Nowadays we can assist, more and more frequently, to videoconferences among people located in different places of the world, just by getting connected to Internet, or listening the radio through a Personal Computer mounting a suitable decoding card. In this case, too considerable amounts of data are to be stored during the transmission and thus fast and high memories are needed.

Without entering particular fields like Computer Science or Telecommunication, in the sphere of common consumer, the employment of memories is foreseen also for common television apparatus which will process data in digital format (DVD).

In car industries, some manufacturers already supply their top class models with a system for satellite navigation, which allows to look for town maps or roads of the place you are travelling in up of a display.

In a next future, it is also foreseen, provided a readjustment of the banking system, that all the credit cards with a magnetic band, will be substitute by intelligent Credit Cards (Smart Cards), having memories that contain all the owner information. That application could be extended to the sanitary field, too, as it has already been done, for the telephone cards, in some European Countries like France and Ireland.

The evidence coming from many applications in every field is to outline that integrated memories have a large employment ranging from the most sophisticated applications to the most common ones.

Obviously in order to satisfy the different demands the memories have had a certain evolution, now they can be divided in two big groups: volatile memories and non-volatile memories.

The RAM (Random Access Memory) belong to the first group, they are used as the main memory in Personal Computers and they have the advantage of a high speed in loading and unloading data as well as a high density of storing (up to date a density of 64 Mbyte is reached). The principal drawback is that this type of memory needs a continuous powering to keep the data, in fact when the power supply is switched off all the data contained in the memory are lost. For this, it is not possible to use it as a mass memory, which has higher densities (Gbyte) to the detriment of a greater slowness. On the other hand it keeps the data also when the system is off.

At the opposite, instead, there are non-volatile memories, which have the prize of keeping the data even if the power supply breaks off. These last ones have had a great evolution, among them we mention the Masked ROM (Read Only Memory) of the first generation. These memories presented the disadvantage that they could not modify the data they contained because they were programmed during the manufacturing process by the fusion of fuses.

Further the OTP (One Time Programmable ROM) were introduced, which, in comparison with the previous ones, had the advantage of having been programmed electrically, even if the impossibility of erasing was still present.

Then the EPROM (Electrically Programmable ROM) came out. They permit an electric writing of the data and their erasing by the exposure of the chip to ultraviolet rays through a window of quartz located on the device package. As we can immediately understand the erasing process is complicated because it is necessary to take the device out of its housing and the erasing time needed is rather long, about twenty minutes.

To finish, the EEPROM (Electrically Erasable Programmable ROM) were introduced, with them both programming and erasing take place electrically reducing the time needed for these operations. The novelty in these memories is the presence of selector transistors for erasing every single cell, obviously to the detriment of a more complex architecture of the control circuits. It is in this situation that Flash memories appeared [1].

As the EEPROM ones, they can be written and erased electrically but this last operation, instead of being done for every cell, should be done for sectors, thus speeds the process up and slims the architecture because it doesn't require the employment of the transistor of selection.

Here we will pay attention to the Flash memory and in particular we will examine its technological characteristics and the working mechanisms of the Flash cell, for a deeper comprehension of writing operations (programming and erasing) and of the relative architectural solution employed to develop these operations, after that the specifications of the device taken into account will be introduced; we will describe the stream followed for the design and the feasibility of a device giving particular attention to the adopted software and to its peculiarities. At the end the programming operations will be examined, reading and erasing of a Flash cell as well as the parts of the circuit interested and the software algorithms executed by the state machine integrated in the chip necessary to develop the over mentioned operations inside, which otherwise should have been done externally by users.

The Flash Memory Cell

In the following text the description of the characteristics of a Flash cell will be given, and in particular the physical processes which are at the base of its working will be underlined.

CMOS memories

A memory cell is an elementary component able to memorise information (bit), it could be found in two physical states corresponding to the logic states "0" and "1". The memories can be divided in two big groups: volatile, the ones which for keeping the data needs to be powered and non-volatile, which can keep the information also when they aren't powered (Flash memories belong to this last group). A first characteristic of the cells of memory can be expressed by its capability of retaining the information compared with the alterability of the memory itself as shown in Figure 1.

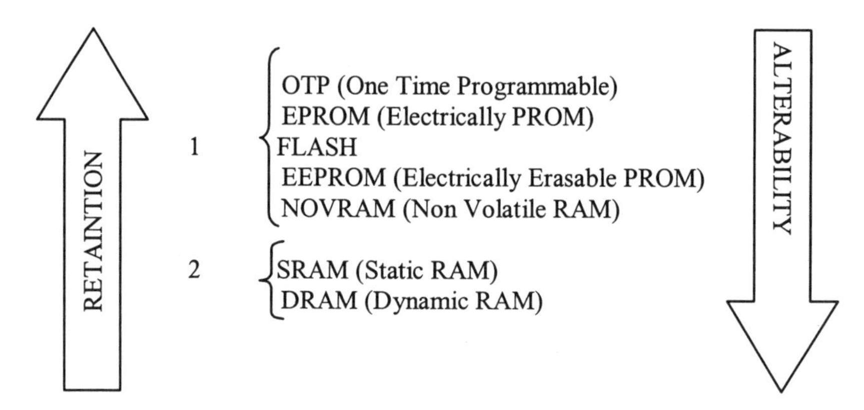

Figure 1: MOS Memories compared with 1) non-volatile 2) volatile

The MOSFET with a floating gate

The fundamental structure of a cell (Figure 2) is composed of a common transistor NMOS, in which between the gate and the bulk a second gate is introduced, completely surrounded by a dielectric, called floating gate [3].

This gate is completely covered and hence not reachable from the outside. Its potential will depend on the charge present on it and also on the capacitive couplings with the other devices of the cell [4] [5].

Furthermore the cell is built in a way that the dominant capacitive coupling of the floating gate should be with the gate reachable from the outside, for this reason it is called control gate.

The dielectric existing between the floating gate and the bulk is called gate oxide or tunnelling oxide (due to the mechanism used to erase the memory) while the one between the two gates is called dielectric inter-poly.

The working principle of the structure, shown in the Figure 2, is the following one: by trapping some electrons in the floating gate we generate on it a potential which shields in part the positive potential presents on the control gate, so we must give a higher potential to this last one to obtain the strong inversion of the channel (see Table 1).

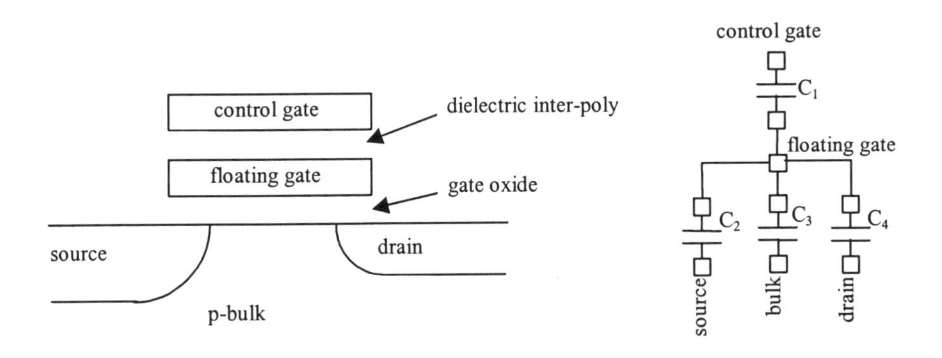

Figure 2: Flash cell a) structure; b) capacitive couplings

In order to distinguish the two possible states of the cell without ambiguity a jump of threshold is requested, for a programmed cell, of at least 3 V, so that the cell threshold voltage results to be not lower than 5V.

Doing so we generate a variation of the MOSFET threshold leading to the distinction between the state when the floating gate doesn't contain any charge, peculiar of a virgin cell or an erased one (represented by a "1" logic) and the state in which the floating gate, being negatively charged, will determine a macroscopic variation of the transistor threshold, peculiar of a programmed cell (represented by "0" logic) [7].

Condition	Threshold voltage	Charge of Floating Gate	Logic State
Virgin/Erased	~2V	None	"1"
Programmed	>5V	Electrons	"0"

Table 1: States of the Flash cell

For this, I/V characteristic in programmed and erased states are parallel with a distance, one from the other, equal to the variation of the threshold voltage (see Figure 3). In reality such an adjustment is no more valid for cells of a memory on which numerous cycles of programming and erasing have been done, as, the cell's charge (its conductance gm) lowers due to the interface states of the bulk oxide which generate a drop of the electrons mobility in the channel.

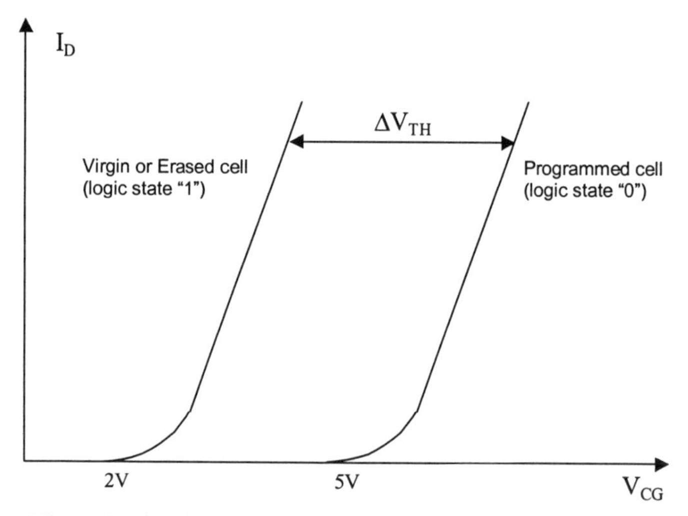

Figure 3: I/V characteristic referring to the programmed and erased cell

Comparison between Flash EPROM and EEPROM memories

In order to avoid any ambiguity we specify that for programming we'll mean the operation by which the electrons are injected into the floating gate (operation of Program: passage from "1" to "0") while for erasing we will indicate the operation leading to the pulling out of the electrons from the floating gate (operation of ERASE: passage from "0" to "1"). For EPROM memories the convention referring to "0" and to "1" logic is opposite.

The Flash EPROM, EEPROM are principally classified following their possibility of being programmed or erased. For this reason we will call EPROM all those devices of non-volatile memory which can be programmed by byte but can't be erased "on board". EEPROM instead will refer to those memories, which can be programmed and erased (on board), by byte. At the end, for Flash memories we will refer to memories which can be programmed by byte but whose erasing can only be possible for sectors.

Even if the EPROM memories can be programmed by byte in a fairly rapid time (<100μs/byte), they have the disadvantage that, in order to be erased, they must be carried out of their housing and then they must be exposed to an ultraviolet radiation (the package they are closed in has a small transparent window to the ultraviolet rays). For this reason the erasing could only be complete and it takes a rather long time (about 20 minutes).

The Flash memories, though have the characteristic they are to be programmable by bytes in a time as long as the EPROM ones do, have the great advantage of being erasable electrically on board, even if only by sectors and not byte by byte. For these kinds of memories the erasing time is greatly reduced to a few seconds. The most flexible one as far as erasing and programming are the EEPROM because they can be erased and programmed by bytes. All this to the detriment of a lost of compactness of the memory matrix owing to the fact that a transistor of selection, which is put in series with the cell itself, is associated to each single cell of memory [6].

Programming and erasing mechanisms

The physical mechanisms allowing programming and erasing of Memory Flash cell, EPROM and EEPROM are basically three: tunnel effect by Fowler Nordheim (FN), injection of hot electrons of the channel (CHEI) and ultraviolet radiation. In the present industrial standard, the EPROM and EEPROM and Flash memories are usually programmed and erased by means of very specific physical mechanisms as shown in Table 2.

The EPROM memories are the only ones whose erasing is obtained by ultraviolet radiation. Such radiation is used in order to give the necessary energy to the electrons present in the floating gate to overcome the potential barrier they are trapped in (photo-emission). At the beginning the electrons start to migrate towards the control gate, till the moment that, besides this stream, an other undesired stream of electrons coming from the bulk and going towards the floating gate starts to overcome the previous one. At "regime" an equilibrium condition is created among the amount of electrons coming from the floating gate and the entering ones, by consequence the cell can't be erased any more.

Condition	EPROM	EEPROM	Flash
Program	CHEI	FN	CHEI
Erase	UV	FN	FN

Table 2: Classification of the EPROM, EEPROM and Flash memories

In the Flash memories for the erasing operation, that is to remove the electrons from the floating gate, it is used the tunnel effect through the gate oxide, called Fowler-Nordheim's [2]. What makes the electrons tunnelling sufficiently probable is the triangular shape of the potential barrier that the electrons must overcome to get out from the floating gate.

In this kind of memories the erasing by bytes can't be obtained because, in a typical structure of a Flash memory matrix the source of all cells are short-circuited each other; the cells belonging to the same column (bite line) have their own drains in common and the cells of the same row (word line) share their own gate. During the erasing operation, consequently, all the cells belonging to a certain sector will be erased simultaneously unless a transistor of selection is put on each row, so in that case it will be possible to erase row by row to the detriment of a great waste in time. The idea at the base of the programming of a Flash cell is the creation of channelling hot electrons in the spatial charge region of drains (area of pinch-off) in order to supply to a certain amount of electrons the necessary energy to overcome the potential barrier of the silicon-oxide interface. In this case, applying a suitable electric field, it's possible to favour the migrations of these electrons towards the floating gate by storing a negative charge on it:

In order to a channelling hot electrons are injected into the floating gate three conditions must take place:

-the electrons must possess a sufficient energy to overcome the potential barrier made by the oxide;

-the momentum must be directed towards the floating gate by means of phonons-electrons or electrons-electrons collisions;

-the electric field in the oxide must be suitable to collect them rather than to reject them against the channel.

The hot electrons receive their own energy from the voltage applied to the cell's drain; they are accelerated along the channel especially in the dumping area of the drain junction where the electric field is much more intense. As soon as a determined threshold of energy is reached, they can overcome the barrier of about 3,2 eV between the bulk and the gate oxide. Part of the energy acquired by the electrons can be lost in different way, like the creation of electron-hole pairs by impact ionisation and the photon emission. As the energy lost due to this physical phenomenon increases with the temperature increasing of the lattice, it's easier to obtain hot electrons at lower working temperatures.

Electrons injected into the gate oxide of a Flash cell as a consequence of the drain and of the channel positive voltage, came back to the bulk, unless an other positive voltage wouldn't be applied to the selection gate in order to attract the electrons towards the floating gate. That's the reason why the floating gate must be polarised positively regarding both the source and the point of the channel where the injection of hot electrons takes place. At the beginning of the injection process (as the electrons stream flows through the oxide at the point where they are hotter and the oxide field is more favourable) the inversion region arrives almost to the drain and in the oxide the electrons field results to be attractive (towards the floating gate). By charging the floating gate, the voltage between this last one and the drain lowers and the area of pinch off of drain moves towards the source; the longitudinal surface field near the drain intensifies and a great amount of hot electrons are produced in the bulk.

	GATE	SOURCE	DRAIN
Read	5V	0V	1V
Program	9V	0V	4.75V
Erase	-9V	4÷6V	Floating

Table3: Applied voltages to the Flash cell

Anyway, in the area where the electrons are hottest (that is near to the drain) the field in the oxide is less favourable for injection (because the energy differences between floating gate and drain lowers step by step as the floating gate loads itself of electrons) and the stream of injected electrons begins to reduce. So the process of injection of electrons limits itself. At the moment in which the floating gate gets completely charged, the energy in the oxide is reduced almost to zero because the field in the oxide become repulsive for the electrons injected in the high field area [18].

As said before the result is that programming and erasing operations take place only if at the control gate, at the source and at the drain must be applied determined values of voltage as described in Table 3 [4].

Simulation and Verification Flow

The "full chip" simulation of the device is executed within several steps through a particular simulation tool. Before describing the simulation flow followed, it is suitable to describe the main instrument used for the simulation, taking into account the features and the advantages offered in comparison with other simulation programs.

The simulation tool: Powermill

The simulation program used for the full chip simulation is Powermill (EPIC), whose feature is no graphic interface with the user, so the instructions are given as text strings through a file. The characteristics of Powermill are [14]:

- ✓ Simulation based on events
- ✓ Technology files based on tables (look up table)
- ✓ Devices described through piecewise linear approximations (PWL)
- ✓ Mixed (analog and digital) simulation optimisation
- ✓ Faster simulation in spite of an accuracy loss

The characteristics of a Spice-like (ELDO) simulation program are [11]:

- Time step based simulation
- Technology file based on mathematical models (model card)
- Devices described through mathematical equations
- Optimised for analog simulations.
- Greater accuracy in spite of a velocity loss

In general Powermill does not substitute ELDO or Spice but they are used together as they have different application fields as a consequence of the listed above characteristics. In fact it is used for:

- ✓ large size circuits (long pattern)
- ✓ functional verifications
- ✓ electrical check up at different accuracy levels
- ✓ full chip simulation.

On the other hand ELDO or Spice can be used for:

- small and medium size circuits (short patterns)
- more accurate electrical check up
- critical sub-circuits simulations.

Powermill simulates the circuit performing a partitioning in entities called stage, each of which interacting with these connected to itself through the events generation process (see Figure 4). Each stage can be evaluated only when an event is generated. The stage are linked to each other through the gates of MOS transistors; MOS transistors connected through the channel belong to the same stage, BJT, inductors and all the other elements have all the terminals into the same stage.

Depending on the case, capacitor placed between two nodes can be converted towards the mass, maintaining both terminals in the same stage.

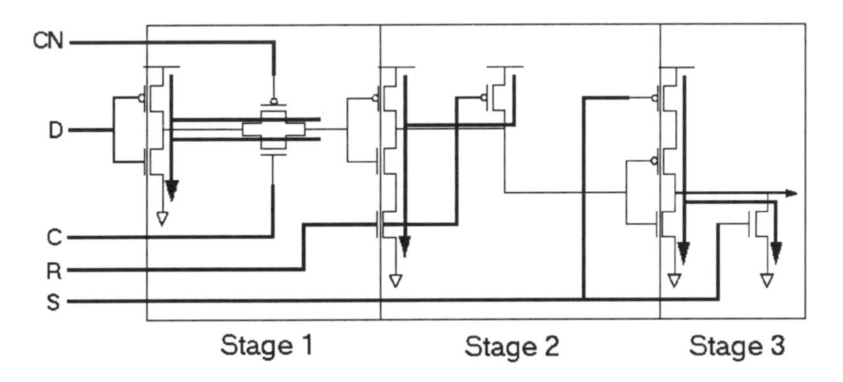

Figure 4: Circuit partitioning performed by Powermill

Powermill allows the simulation of a circuit at different accuracy levels, which can also be modified only in one of the sub-blocks. The simulation velocity changes with the accuracy level chosen.

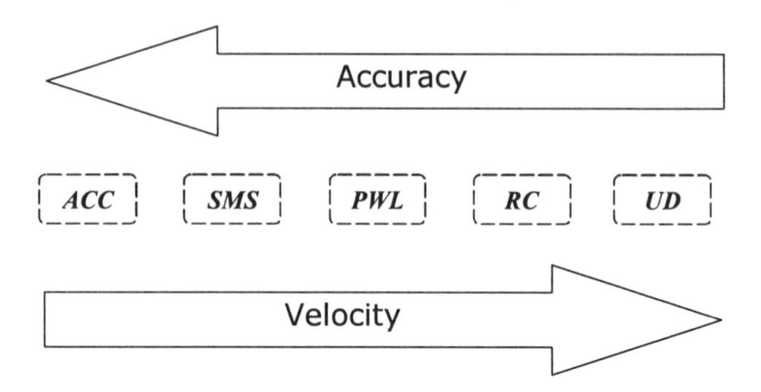

Figure 5: Level of accuracy

To each level of accuracy corresponds a different level of description of the devices presents in the circuit.

PWL: is the default modality used by the tool. The devices are described through a fixed capacity model; there is no bulk current simulation: gate capacitors of the MOS are supposed to be towards the mass allowing the partitioning of the circuit into de-coupled stages; only when an event is generated at one of its gates a new value is given to the stage (revealed); nodes can be both analog and digital.

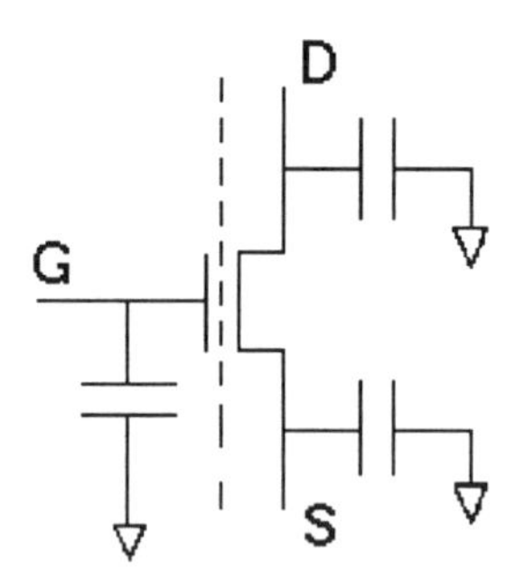

Figure 6: Partitioning in modality PWL

SMS: in the modality more accuracy is used to describe the devices; capacitors vary in accordance with voltage; the gate capacity of the MOS is distributed between drain and source in conformity with the Miller effect; partitioning is maintained but stages are synchronised through an algorithm (SMS = synchronous matrix solver); nodes can be digital and analog.

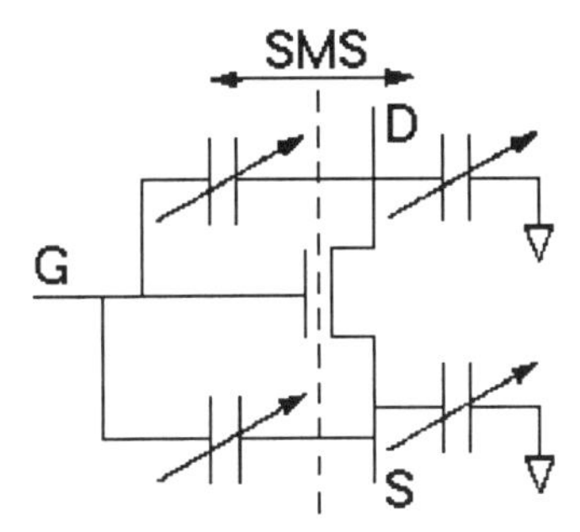

Figure 7: Partitioning in modality SMS

ACC: this modality allows a level of accuracy comparable to the ELDO or SPICE ones, it is based on the same variable capacitors model used in the SMS mobility; eventual positions are gathered into a single stage; all nodes are analog, using this modality it requires a generation of large stages, which slows the simulation.

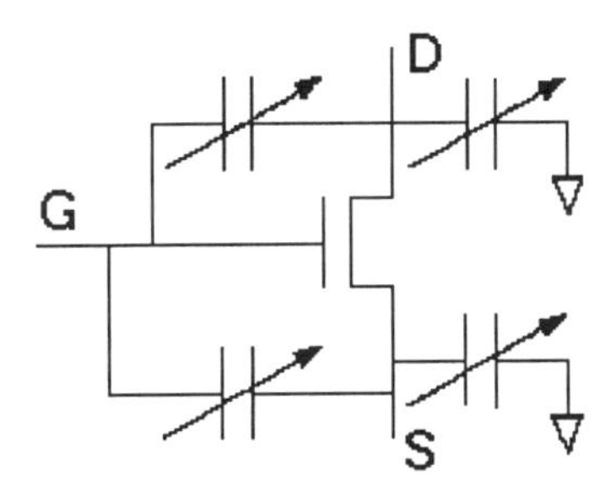

Figure 8: Partitioning in modality ACC

RC: in this modality, models simplified with resistors and capacitors (switch level) are used to describe the devices; it can only perform simulation on digital circuits.

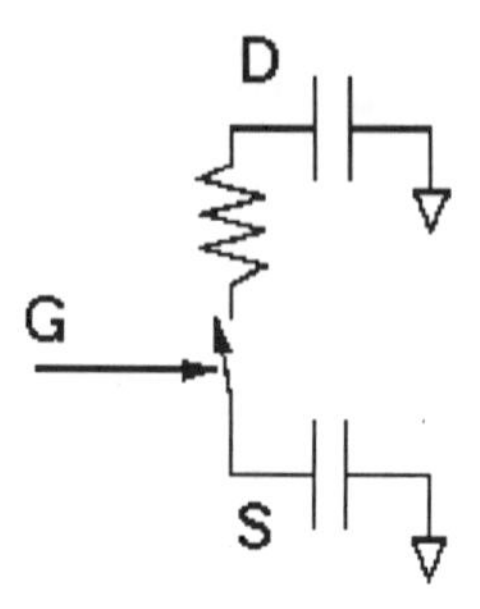

Figure 9: Partitioning in modality RC

UD: (unit delay) it is a modality which describes the devices through simplified fixed delays models (gate level); it simulates only digital circuits.

To throw a simulation with Powermill it is necessary to write down a text file "script" indicating the configuration file, the netlist and the technology. Finally the program will produce an output file (out) including data of the simulated circuit (see Figure 10).

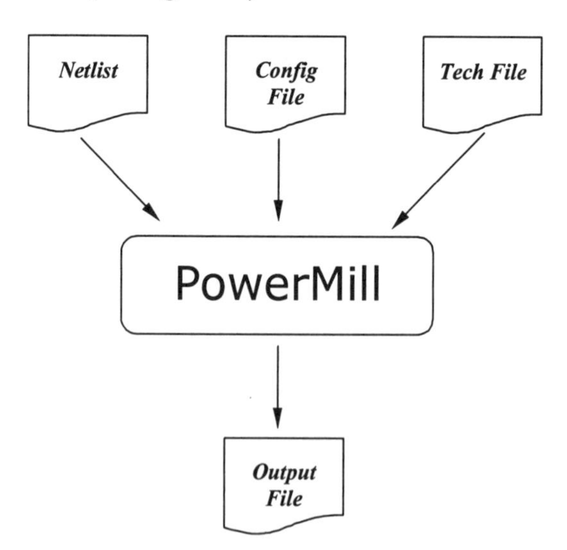

Figure 10: Powermill I/O flow

The netlist includes information about the topology of the circuit and the stimuli to be applied in the entry. Powermill accepts netlist in the entry file compatible with different simulators (ELDO, Spice, Verilog...)

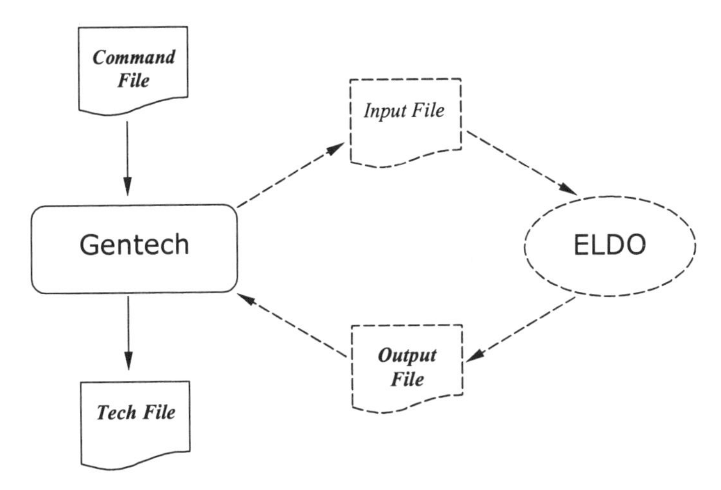

Figure 11: Generation of tech files

Through the configuration file, instructions can be given to the simulator, like analysis to be performed, waveforms in output, partitioning control and level of accuracy.

The technology file includes all the tables describing the circuits of the device. The file can be generated with a program (see Figure 11), Gentech, [12] which gives results in tables form ready to be used by Powermill as input file. The table form is obtained by ELDO simulations. Usually those files are available for every technology therefore there is no need to generate them for every simulation.
The output file of Powermill contains the data regarding the signals to be visualised (in a graphic shape) through compatible viewer.

Simulation flow and verification

The flow chart shown in the Figure 12 describes the steps followed to perform the "full chip" simulations of the device. The Figure also shows the flow regarding the simulation check and comparison.

The file containing the topology of the circuit, the netlist, is obtained through a sub-program of the tool OPUS (cadence), the "Netlister", which translate the scheme of the whole device into a text-file. The OPUS tool is used by designers to design the circuit and create the database necessary to perform the simulations.

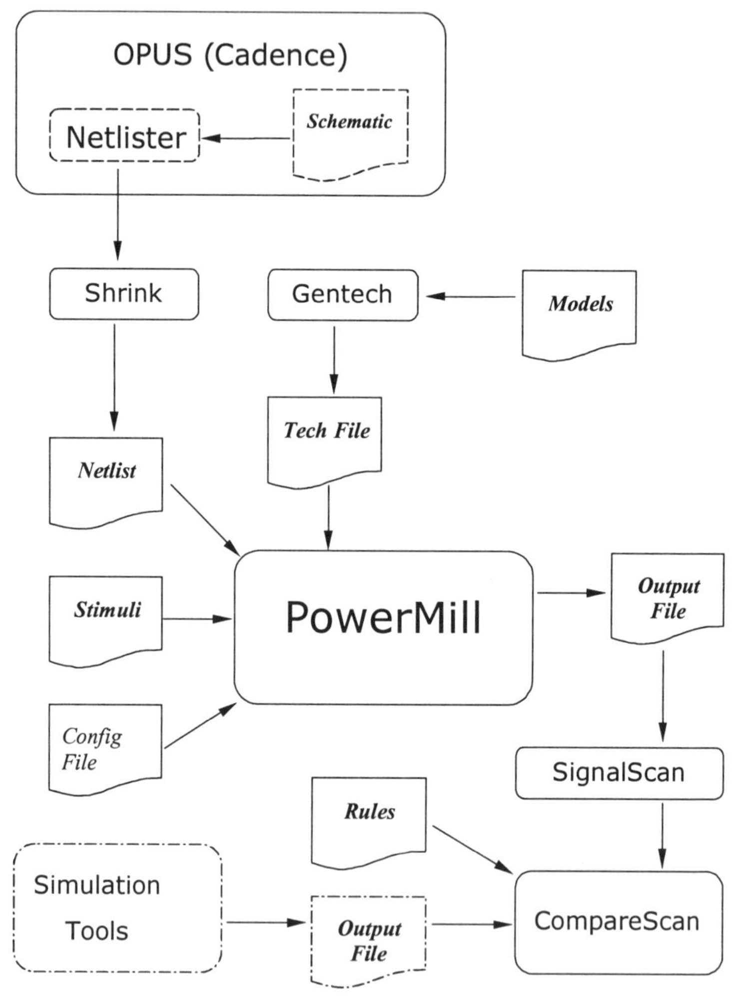

Figure 12: Simulation and verify flow

On the netlist, obtained with Opus, it is necessary to perform the "scaling" of all the devices included in the chip depending on the technology used

(0,35 μm), such operation is performed by a program written in Unix, the Shrink, which gives back as output the new netlist.

Unlike what previously described, it is preferable to write down a stimuli file rather than include it in the netlist, as the netlist is normally a large file, therefore hardly to handle through text editors. The stimuli file describes the signals to be applied on input pins of the chip for the various simulations; such signals are described through a simple language similar to Pascal.

Once the netlist is ready, the technology file has been indicated, the configuration and the stimuli files have been written, the simulation of the whole chip can be started with Powermill.

Once the simulation is done Powermill generates the output file (.out) which can be graphically displayed through the Signalscan tool (DAI). This program, besides the plot of the output waveforms, creates a database where all the signals obtained through the simulation are stored.

The Comparescan (DAI) tool can compare, by following fixed rules, the database of simulations performed in different ambiences or simulations performed in the same ambient but on different devices or on different version of the same device. The rules can be set through a file (Rules) where comparison tolerances can be specified.

Through Comparescan it is possible to perform the automatic check.

Electrical and Functional Characteristics of the Device

Now the electrical and functional features of a specific device will be examined. Successively the architecture of the device will be studied in details by carrying out some simulations that verify its correct performance.

General Characteristics

The device taken into consideration is the M29W160BT-B; it is a non-volatile Flash memory of 16 Mbit, at a single low voltage supply [13].

The sequence of letters and numbers serves to identify the product type, in fact M29, indicates a device with single only voltage reference ranging between 2.7 and 3.6 V; 160B indicates that it is referred to a memory of 16 Megabit with the Boot Block that could be found physically in the higher (Top level) or lower (Bottom level) area of the matrix as it could be seen from the remaining characters.

In reality the mark of the device is characterised by further characters identifying speed, type of package and temperature range. In the following table the main characteristics of the examined device are summed up.

Single supply voltage	2.7-3.6 V program and erase
Access time	70 ns
Programming time	10 μs Byte/Word
Memory Block	35 (1 boot block, 2 parametric sectors, 32 main sectors)
Program/Erase controller	Internally program algorithm byte/word Internally erase algorithm sector/matrix
Erase suspend and resume	Read or Erase suspend and then resume
Temporary Block unprotection mode	Temporary Block unlock mode
Security memory block	Protection of sector
Low power consumption	Automatic standby
Program/erase cycle	100.000 cycle of operation for each sector
Data retention	20 year with defect of 1ppm/anno

Table 4: Characteristics of the device

The device is made of a cell matrix where information are stored up and of a kind of control circuit permitting to carry out not only reading and writing operations but also other ones as shown in the Table 4. When the memory switches on the device pre-sets automatically in reading manner. The matrix is divided into 35 sectors (addressable from the address 000000 to 1FFFFF one) which could be erased simultaneously or separately in accordance with the needing.

Sectors could farther be distinguished in 31 blocks of 64 Kbyte each one where the user can memorise data and other 4 blocks of 32, 16, 8, 8 Kbyte forming the so called Boot Block. The block of 16 Kbyte can be used to memorise the initialisation code of the microprocessor, those of 8 Kbyte could be employed to store parameters and the remaining 32 Kbyte block could serve as further place to store data. These four blocks could be at the top of the matrix so in this case it is said *top* configuration, or it could be in the lower part, in this case it is said *bottom* configuration.

Each block could be protected singularly to prevent casual errors like an erasing or an undesired programming.

The device has among its control circuit an interface memory-microprocessor (the so-called Command User Interface) in which the necessary instructions to perform the operations of reading and writing are written. The over mentioned instructions decoded are sent to a micro-controller integrated in the chip which verifies that the operations will be performed exactly underlying the result or eventual error conditions.

The instructions written on the CUI are sent to the controller to start the inner algorithms of programming and erasing. The remarkable advantage of this architecture is to avoid the possibility for the user to carry out these operations from the outside. In the following Figure 13 are shown the input and the output of the examined device.

The signals, *chip enable*, *output enable* and *write enable* control the operations between bus and memory. They allow a direct link between microprocessor and memory without any further control logic.

Detailed description of signals.

Here is a detailed description of each signal shown in Figure 13:

Address input (A0-A19): The address bus is used to select the memory cells in the matrix during reading and writing operation. During the last one the A0-A19 control the signals sent to CUI of state machine.

Data Inputs/Outputs (DQ0-DQ7): DQ0-DQ7 stands for data bus. The bus gives back the stored information in the memory cells chosen during the reading operation. During the writing operation they represent the instructions sent to CUI of the state machine.

Data Inputs/Outputs (DQ8-DQ14): DQ8-DQ14 give back the data stored in the memory cells chosen during reading operation if the signal "byte" is low (V_{IL}). When "byte" is high (V_{IH}) the pins in output are placed in a high impedance. During the writing operation the register of instruction doesn't use these pins, so reading the status register these bits are ignored.

Date Inputs/Outputs or Address Input (DQ15A-1): When "byte" is low (V_{IL}) this pin acts as an input and output data in such a way allowing a reading for 16 (DQ0-DQ15) together with the other pins. When the "byte" is high (V_{IH}) the reading is for 8; in this case only the pins DQ0-DQ7 are necessary, while the pin is used to an address pin. In this case DQ15A-1 low selects the least significant bit (LSB) in the word we find in pins A0-A19. DQ15-1 high selects the most significant one, instead.

Chip Enable (E): The *chip enable* allows the memory to do reading and writing process. When this pin is high the memory is not enable for any kind of operation.

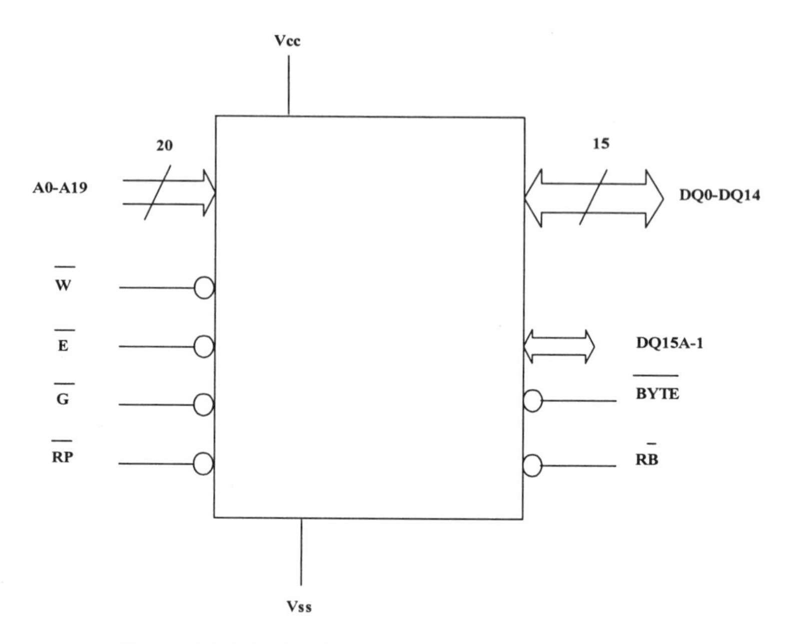

Figure 13: Block schema of the device M29W160BT

Output Enable (G): The *output enable* enables the memory to give back the data on the output pins during the reading operation. This signal is working when it is low.

Write Enable (W): The *writing enable* enables the CUI of the memory for a writing operation. This signal is working when is low.

Reset/Block Temporary Unprotect (RP): This signal can be used to reset the hardware of the memory or to unprotect momentaneously all the sectors protected before. Instead when this signal gets to a high level (V_{IH}) the memory is ready for a reading operation and then for writing one. Carrying the signal to V_{ID} all the sectors protected won't be protected any more and the writing operation becomes possible.

Ready/Busy Output (RB): This pin is a terminal of an open drain that allows to see when the reading operation is enabled. During the Read mode, the Auto-select mode and the Erase Suspend mode, the pin is in high impedance. After a reset of the hardware it's not possible to do any writing or reading operations if the pin is not in high impedance. When the operations of programming and erasing start the pin runs to a low level (V_{OL}) and it remains there during reading operation, too. A low value of this pin marks that the memory is busy doing an operation. The choice of connecting the pin to the open drain is due to the fact that in such a way it is possible to link that pin to the same pin of other memories to a "unique pull-up resistor.

A0-A19	Address inputs
DQ0-DQ7	Data inputs/outputs
DQ8-DQ14	Data inputs/outputs
DQ15A-1	Data in/out or Address input
E	Chip enable
G	Output enable
W	Write enable
RP	Reset/block temporary unprotect
RB	Ready/Busy output
Byte	Byte/word organisation select
Vcc	Voltage supply
Vss	Ground

Table 5: Signals description

Byte/Word Organisation Select (BYTE): This pin is used for changing the read modality 8 bit to the 16 bit one of the data bus. When "byte" is high (V_{IH}) the memory is working in modality 8 bit, when the pin is low (V_{IL}) the memory is working in 16 bit.

Vcc Supply Voltage: The supply voltage is connected to this pin. The CUI is disabled when the supply voltage is lower than a voltage called Lockout Voltage (V_{LKO}). Going down this voltage it prevents undesired operations of writing and this happens during the operations of switch on or switch off of the memory avoiding the possibility that the contained data should be changed in the cells. If the state machine has started the inner programming and erasing algorithms, the operations are interrupted considering not valid the content of altered cells. A capacitance of 0.1 μF must be connected between voltage supply and ground in order to avoid passage of current.

Vss Ground: This is the voltage reference.

Bus operations

There are 5 operations which involve the bus: Bus Read, Bus Write, Output Disable, Standby and Automatic Standby. As we can see, in Table 6 and 7 the signals involved in such operations and their status are put in evidence.

Operation	E	G	W	DQ15A-1, A0-A19	DQ14-DQ8	DQ7-DQ0
Bus read	V_{IL}	V_{IL}	V_{IH}	Cell Address	Hi-Z	Data Output
Bus write	V_{IL}	V_{IH}	V_{IL}	Command Addr.	Hi-Z	Data Input
Output Disable	X	V_{IH}	V_{IH}	X	Hi-Z	Hi-Z
Standby	V_{IH}	X	X	X	Hi-Z	Hi-Z
Read-Man. Code	V_{IL}	V_{IL}	V_{IH}	A0=V_{IL}, A1=V_{IL} A9=V_{ID},other V_{IL} or V_{IH}	Hi-Z	20h
Read Dev. Code	V_{IL}	V_{IL}	V_{IH}	A0=V_{IH}, A1=V_{IL} A9=V_{ID},other V_{IL} or V_{IH}	Hi-Z	C4h(M29W160BT) 49h(M29W160BB)

Table 6: Bus Operations, Byte=V_{IL}, X=V_{IL} or V_{IH}

Operation	E	G	W	A0-A19	DQ15A-1, DQ14-DQ0
Bus read	V_{IL}	V_{IL}	V_{IH}	Cell Address	Data Output
Bus write	V_{IL}	V_{IH}	V_{IL}	Com. Address.	Data Input
Output Disable	X	V_{IH}	V_{IH}	X	Hi-Z
Standby	V_{IH}	X	X	X	Hi-Z
Read-Manual. Code	V_{IL}	V_{IL}	V_{IH}	A0=V_{IL}, A1=V_{IL} A9=V_{ID}, other V_{IL} or V_{IH}	0020h
Read Dev. Code	V_{IL}	V_{IL}	V_{IH}	A0=V_{IH}, A1=V_{IL} A9=V_{ID},other V_{IL} or V_{IH}	22C4h(M29W160BT) 2249h(M29W160BB)

Table 7: Bus Operations, Byte=V_{IH}, X=V_{IL} or V_{IH}

Glitches of 5 ns or less, on the *chip enable* and on the *write enable* are ignored by the memory; thus they don't influence the operations. More in detail the five operations are:

Bus Read: As previously said the reading operations allow the reading of cells contents or the reading of a Command User Interface register specifying the address bus. This operation is possible when the *chip enable* and the *output enable* are on V_{IL} while the *write enable* is at a high value V_{HL}. In this way pins DQ0-DQ7 give back the data in accordance with what we can see from the Table and by the following specified schedules.

Bus Write: The operation allows the writing of the instruction on the CUI. A correct writing operation requires the setting of the address bus level with the falling low edge of the *chip enable* or the *write enable* depending on which one arrives first. On the other hand the CUI sets the data bus on the rise edge of the *chip enable* or of *write enable* depending on which of the two arrives first. Obviously the *output enable* will remain at the V_{IH} level during the whole writing operation.

Output Disable: The pins DQ0-DQ7 are in a high impedance state when G is high (V_{IH}).

Standby: When E is high the memory is in standby and the output pins are at high impedance. To reduce the dissipation E is set to the value Vcc $+\Delta V$ (being $\Delta V = \pm\ 0.2$). The currents and voltages values of this operation are reported in Table 8.

Symbol	Parameter	Test Condition	Minimum	Maximum	Unit
I_{LI}	Input-leakage current	$0V \leq V_{IN} \leq V_{CC}$		+-1	μA
I_{LO}	Output-leakeage current	$0V \leq V_{OUT} \leq V_{CC}$		+-1	μA
I_{CC1}	Supply current (Read)	$E=V_{IL}$, $G=V_{IH}$, f=6MHz		10	mA
I_{CC2}	Supply current (Standby)	$E=V_{CC}$+-0.2V		100	μA
I_{CC3}	Supply current (Program/Erase)	Program/Erase Controller active		20	mA
V_{IL}	In-low voltage		-0.5	-0.8	V
V_{IH}	In-high voltage		0.7 V_{CC}	V_{CC}+0.3	V
V_{OL}	Out-low voltage	I_{OL}=1.8mA		0.45	V
V_{OH}	Out-high voltage	I_{OH}=-100μA	V_{CC}-0.4		V
V_{ID}	Identification V		11.5	12.5	V
I_{ID}	Identification C	A9=V_{ID}		100	μA
V_{LKO}	Program/Erase Lockout Voltage		1.8	2.3	V

Table 8: Characteristic DC

Automatic Standby: If the CMOS levels (Vcc = –0.2) are used to drive the bus and the bus stay inactive for 150 ns or more, the memory is automatically put in standby and the values of the current are those shown in the Table 8 and the data bus gives back values if the operation is reading one.

Electronic Signature: the bus can perform other operations during the programming phase of the device, but not during usual operations performed by the memory. One of such operations is to read the manufacturer code and the device code, which are useful to identify the memory.

Block Protection and Block Unprotection: the second special operation is the one which allows the protection sectors against accidental writing and erasing, and unprotection of sectors in case of need to change protected data.

Command	Lenght	Bus Write Operations											
		1st		2nd		3rd		4th		5th		6th	
		Addr	Data	Addr	Data	Addr	Data	Addr	Data	Addr	Data	Addr	Data
Read/Reset	1	X	F0										
	3	555	AA	2AA	55	X	F0						
Auto select	3	555	AA	2AA	55	555	90						
Program	4	555	AA	2AA	55	555	A0	PA	PD				
Unlock Bypass	3	555	AA	2AA	55	555	20						
Unlock Bypass Program	2	X	A0	PA	PD								
Unlock Bypass Reset	2	X	90	X	00								
Chip Erase	6	555	AA	2AA	55	555	80	555	AA	2AA	55	555	10
Block Erase	6+	555	AA	2AA	55	555	80	555	AA	2AA	55	BA	30
Erase Suspend	1	X	B0										
Erase Resume	1	X	30										
Security Data	1	X	98										

Table 9: Instructions in 16 bit modality, Byte=V_{IL}

Command	Lenght	Bus Write Operations											
		1st		2nd		3rd		4th		5th		6th	
		Addr	Data	Addr	Data	Addr	Data	Addr	Data	Addr	Data	Addr	Data
Read/Reset	1	X	F0										
	3	AAA	AA	555	55	X	F0						
Auto select	3	AAA	AA	555	55	AAA	90						
Program	4	AAA	AA	555	55	AAA	A0	PA	PD				
Unlock Bypass	3	AAA	AA	555	55	AAA	20						
Unlock Bypass Program	2	X	A0	PA	PD								
Unlock Bypass Reset	2	X	90	X	00								
Chip Erase	6	AAA	AA	555	55	AAA	80	AAA	AA	555	55	AAA	10
Block Erase	6+	AAA	AA	555	55	AAA	80	AAA	AA	555	55	BA	30
Erase Suspend	1	X	B0										
Erase Resume	1	X	30										
Security Data	1	X	98										

Table 10: Instructions in 8 bit modality, Byte=V_{IH}

Command Interface

Each writing on the memory consists of a list of instructions that the Command Interface (CUI) interprets. Such list of instructions is executed to guarantee the correct transfer of the data. A failure while executing such instructions brings back the memory to the reading modality. The addresses used during various operations change depending on the setting of the memory in modality 8 bit or 16 bit. Table 9 and 10 allow a deeper understanding of the various operations.

It is now worth to proceed to a deeper analysis of the instructions:

Read/Reset Command: This instruction sets the memory in read modality and resets errors occurred in the status register. As can be seen in the tables this operation can be performed within 1 or 3 clock cycles involving the data bus and the address bus once or three times. If this instruction is performed during an operation of erasing it will take 10 μs before such operation could be interrupted. During this period output data can not be read. Thus this instructions has, as a consequence, to leave data stored in the memory unchanged.

AutoSelect Command: This instruction is used to read the Manufacturer Code, the Device Code and the Block Protection Status. In this case three operations are required on the bus and the memory remains in such modality until a new instruction is given. As can be seen in Table 7 putting A0 = V_{IL}, A1 = V_{IL} it is possible to read the Manufacturer Code which is 0020h for the ST. To read the Device Code it is enough to put A0 = V_{IH} and A1 = V_{IL} while to read the protection status of the sector it's enough to put A0 = V_{IL} and A1 = V_{IH} specifying the address of the sector through A12-A19. Any value can be given to the remaining set of bit. If the addressed sector is protected DQ0-DQ7 give back 01h, otherwise 00h.

Program Command: The program instruction is used to store data in the array of the memory specified by the address. This instruction requires four bus cycles, as can be seen in the Tables 9 and 10, and during the fourth cycle address and the data are stored in the state machine, which starts the program/erase algorithm. If the address refers to a protected sector the program instruction is ignored and the data remains unchanged. In this case the status register can not be read and no error condition is reported. During the program operation all the other instructions are ignored, avoiding any interrupt of the programming. The reading operation, performed during the programming operation, results in the report of the content of the register in the output pins. If there is no error during the programming, the memory is automatically set in read modality, otherwise the content of the register is reported in output and the Read/Reset instruction must be issued to reset the error condition and return to read modality. The program instruction cannot change to 1 a bit set to 0. To perform this operation it is necessary to give an Erase instruction that sets to 1 the bits of one sector or of the whole matrix.

In the following Table are shown typical endurance required to perform same instructions.

Parameter	Min.	Typ.	Typ.-After 100K W/E cycles	Max.	Unit
Chip Erase (all bit in matrix. are set to 0)		10	10		sec
Chip Erase		22	22		sec
Block-Erase (64kbytes)		0.8	0.8	t.b.d	sec
Program (byte or word)		10	10		μs
Chip-Program (byte by byte)		22	22		sec
Chip-Program (word by word)		11	11		sec
Program/Erase Cycles (per-Block)	100.000				cycles

Table 11: Program, Erase Time, Erase Endurance Cycles (T_A = 0 to 70 °C)

Unlock Bypass Command: This instruction is used together with the Unlock Bypass Program. When the access time is long it is possible to speed up the system using this instruction which requires three bus operations. After the performing of the instruction the memory executes only the Unlock Bypass Program or the Unlock Bypass Reset. In this case it is possible to read the memory if it is set in read modality.

Unlock Bypass Program Command: This instruction is used to program an address in the memory. It requires two bus operations during which the address and the data are memorised in the state machine, which checks the correct execution of the Program/Erase Algorithm. This command has the same effect of the program instruction. As can be guessed, a protected sector cannot be programmed and the operation cannot be interrupted. At the end of the operation the content of the status register is read, so that if an error has occurred it can be reset with the Read/Reset instruction, which leaves the memory in Unlock Bypass modality.

Unlock Bypass Reset Command: This instruction sets the memory in Read/Reset modality after the execution of the Unlock Bypass instruction. In this case only two bus operations are required.

Chip Erase: Through this instruction it is possible to erase the whole memory matrix. Six bus cycles are required to enable the internal erase algorithm. All the protected blocks are ignored while the others are erased. If every block is protected the erasing algorithm starts but finishes within a 100 μs time leaving data unchanged and no error condition is reported. During the erase operation all the other instructions are ignored preventing it from being interrupted. A bus read operation gives back the contents of the status register. At the end of operation the memory is set in read modality unless errors occur. In this case the memory gives as output the content of the status register. To reset errors the Read/Reset instruction must be given and the memory is set back to the read modality. The chip erase sets the bits of unprotected cells to 1.

Block Erase Command: The instruction, at the opposite of the previous one, allows the erasing of a list of blocks. Six bus cycles are required to select the first sector of the list. Any other sector of the list can be erased performing again the six bus cycle. The instruction starts the erase algorithm 50 μs after the last writing operation. Once the algorithm is started there is no possibility of changing the specified sector. The selection of a different block must take place with 50 μs. The status register can be read after the six bus cycles. Only unprotected blocks are erased. In this case too, if every sector is protected the erase algorithm starts but it is stopped with 100 μs without that any error condition is reported. During the operation any other instruction is ignored except the Erase/Suspend and the Read/Reset. The bus reading performed during the operation has the result of reporting as output the content of the status register. At the end of the operation the memory goes back to the read modality unless an error has occurred. In this case the memory gives back as output the content of the status register and it is necessary to give the Read/Reset instruction to reset the errors and set the memory to the read modality. The instruction puts the bits of unprotected sector to 1.

Erase Suspend Command: The instruction, as previously said, allows the temporary suspension of the Block Erase Command and sets the memory to the read modality. The instruction requires a single cycle of the bus. The erase algorithm is suspended after 15 μs the instruction is given. If the instruction is given while the memory is waiting for a new sector specified before resuming the Block Erase Command, then the Erase Suspend immediately acts without any more waiting, subsequently the Erase Resume instruction is immediately resumed, and after that it is no longer possible to specify a different sector. During the Erase Suspend it is possible to read and program cells of sectors that has not been selected for the erasing. The result of a reading of an erased sector gives back the content of the status register.

Erase Resume Command: The instruction allows the restart of Program/Erase Algorithm after having performed the Erase Suspend instruction. An erasing can be suspended and resumed more than once.

Security Data Command: It is used to read the security sector of memory. This one is a 256 words sector. Actually a user can require that a single security code be pre-programmed by the company producing the chip. A single bus cycle is required to perform the instruction, which imposes the reading of this block only till another instruction is given. The addresses of the security block are reported in Table 12.

Size (Word)	Address Range(x8)	Address Range (x16)
256	000000h-0001FFh	000000h-0000FFh

Table 12: Addresses of the security block of the memory

Status Register

The reading of the bus implies the reading of the content of the status register during the various operations of programming and erasing. The register is also read during an Erase Suspend when the access to the address of an erased block is reached. The bits in the status register are resumed in Table 13. The function of the bits of the status register is, more in detail:

Operation	Address	DQ7	DQ6	DQ5	DQ3	DQ2	RB
Program	Prog. Addr.	DQ7 negate	Toggle	0	-	-	0
Program-during Erase Suspend	Prog. Addr.	DQ7 negate	Toggle	0	-	-	0
Program Error	Prog. Addr.	DQ7 negate	Toggle	1	-	-	0
Chip Erase	Any Addr.	0	Toggle	0	1	Toggle	0
Block-Erase before Timeout	Block Addr.	0	Toggle	0	0	Toggle	0
Block Erase	Block Addr.	0	Toggle	0	1	Toggle	0
Erase Suspend	Erasing Block	1	No Toggle	0	1	Toggle	1
Erase Suspend	Non-Erasing Block		Data read as normal				1
Erase Error	Good-Block Address	0	Toggle	1	1	No Toggle	0
	Faulty-Block Address	0	Toggle	1	1	Toggle	0

Table 13: The bits of the status register

Data Polling Bit (DQ7): This bit is used to identify whether the Program/Erase Algorithm has been successfully performed or if there is an Erase Suspend going on. As it can be seen from Table 10, during the programming the pin, which the DQ7 is referring to, is at 0. At the opposite, during an Erase Suspend the DQ7 is at 1 if a reading of the bus of an erased sector is being performed. If the state machine interrupts the erasing algorithm than the DQ7 changes value from 0 to 1.

Toggle Bit (DQ6): The DQ6 too is used to check whether the Program/Erase Algorithm has successfully completed the operation or the state machine is responding to an Erase Suspend. The DQ6 is in toggle state during the reading of the Status Register. During the Program/Erase operations the DQ6 continually changes its value from 0 to 1 in accordance with subsequent readings of the bus at the address of the programmed or erased sector. During the Erase Suspend the DQ6 will give back a value to the corresponding output pin when a cell of the sector to be erased is addressed. The DQ6 will no longer be in toggle state when the state machine suspends the erase operation.

Error Bit (DQ5): The DQ5 is used to check whether an error has occurred during the execution of the internal Program/Erase Algorithms. The error bit is at 1 when the Program operation, the Block Erase operation and the Chip Erase operation failed. In this case the Read/Reset instruction is required before performing any other instruction.

Erase Timer Bit (DQ3): This bit is used to identify the starting time of the erase algorithm that follows the Block Erase. When the state machine starts the algorithm the DQ3 runs from 0 to 1. Only when the DQ3 is at 0, the address of the another block to be erased can to be specified in the CUI.

Alternative Toggle Bit (DQ2): The DQ2 is used to check the state machine during the erase operation. During the Chip Erase operation and the Block Erase operation the toggle bit turns alternatively from 0 to 1 with subsequent readings of the bus at the address of the erased block. During the Erase Suspend too, the DQ2 alternatively turns from 0 to 1 and a reading of the bus at the addresses of the blocks that have not been erased gives back as output the data stored in the cells of such blocks. After an erase operation, which implies the setting of the DQ5, the DQ2 is used to identify the block or the blocks that caused the error. In this case this bit alternatively turns from 0 to 1 with subsequent readings of the bus at the addresses of the blocks that have not been correctly erased and caused the error. Instead, the bit does not change value when the addressed block has been erased correctly.

Read Operation

Now the reading operation will be examined in details. As previously said, it consist of a phase in which instructions are sent to the Command Uset Interface and a phase in which the internal state machine controls the ongoing reading operation, consisting in the decoding of row and column and in the reading of the content of the cells in the sector at the corresponding address, through a circuit called Sense Amplifier. Each of these phases will be described highlighting the main passages of the operation.

The Finite State Machine

As it has been observed, to perform one of the operations allowed in the memory, it is necessary to give a list of instructions to the CUI. The list of instructions consists in a fixed configurations sent to the address bus and to the data bus. The CUI is a part of the internally integrated state machine, which controls the correct execution of the inner algorithms that started as a consequence of the given instruction.

So the FSM (Finite State Machine) performs the algorithms that has been resumed in the blocks scheme reported in the Figure 14 [8] [2].

It is possible to observe in the diagram the *isfree* condition regarding to the byte program, which is used to check whether the addressed byte is part of the protected sector.

Numbers on the side of the links are the boolean states of true-false. In this way, the users send the list of instructions without performing any kind of control that instead is performed by the State Machine (such as, for instance, checking the threshold of the cells or performing a soft programming of the cells before the erase operation to prevent the depletion of some of them).

The information advice the user about the state of the FSM and eventually occurred errors are reported on the external pins.

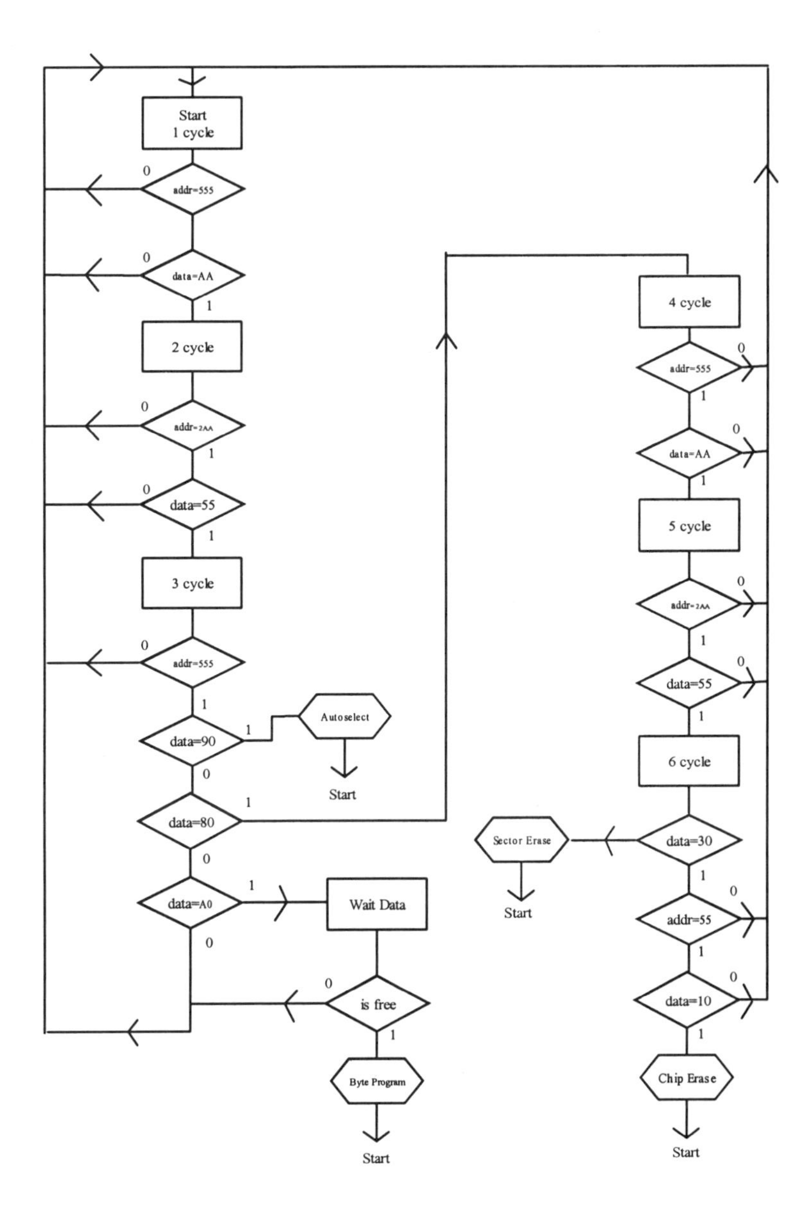

Figure 14: Finite State Machine Algorithm: 16 bit command mode (Byte=V_{IH})

Blocks Structure of the Device

As can be guessed from what has been described previously, the device M29W160BT is really complex and it must include all the circuits needed to read, erase and program the cells of the memory. Such circuits must generate into the chip suitable voltages (charge pumps) and adjust them (voltage regulators).

Circuits which can perform operations as fast as possible and can execute a correct timing between different operations in order to prevent errors, are required. Besides there are tightening specifications on the memory cell which cannot be heavily charged (output buffers, which introduce as low noise as possible on the supply, have then been provided) and cannot be stressed with unsuitable voltage mainly when the cell has not been selected for the required operation. Moreover if it is taken into consideration that the memory matrix consist of 16 millions of cells, it comes as consequence the need to allow a correct decoding of row and column to identify, for instance, the word to be read through a structure called sense amplifier.

Besides, as the integration process does not have a yield of 100% on the silicon slice, it is used an exceeding numbered cells called "redundancy cell", which allows the substitution of the faulty ones without compromising the functionality of the whole device. Circuits, which can select such cells that are part of additional columns and also latches where it is possible to memorise the address of the column to be redounded, are required.

Naturally all these things are transparent to the user who cannot see what happens within the device. In the Figure 15 a simplified block scheme of the layout (floor plan) of the chip is shown so that it is possible to recognise some of the mentioned circuits. The device consists of a left and right half-matrix (in the Figure 15 it is shown the blocks scheme of one of the half-matrix). Each half-matrix is equivalent to an 8 Mbit matrix. The left half-matrix has 16 sectors of 64 Kbyte each one consisting of 256 rows (word lines) and 2048 columns (bit lines), each sector is then divided in two half-sectors each one consisting of 1024 columns. The right half-matrix, which is a copy of the left one, consists of 15 sectors of the matrix, besides the Boot Block and the CFI, so that there are totally 35 sectors, as it was said previously.

Moreover the first quarter of the columns corresponds to the first of the eight outputs, the second quarter to the second output and so on. Each word, thus, is identified on a single row selecting eight columns at the same time.

Columns belonging to each quarter of half-sector have the same address and are selected contemporary.

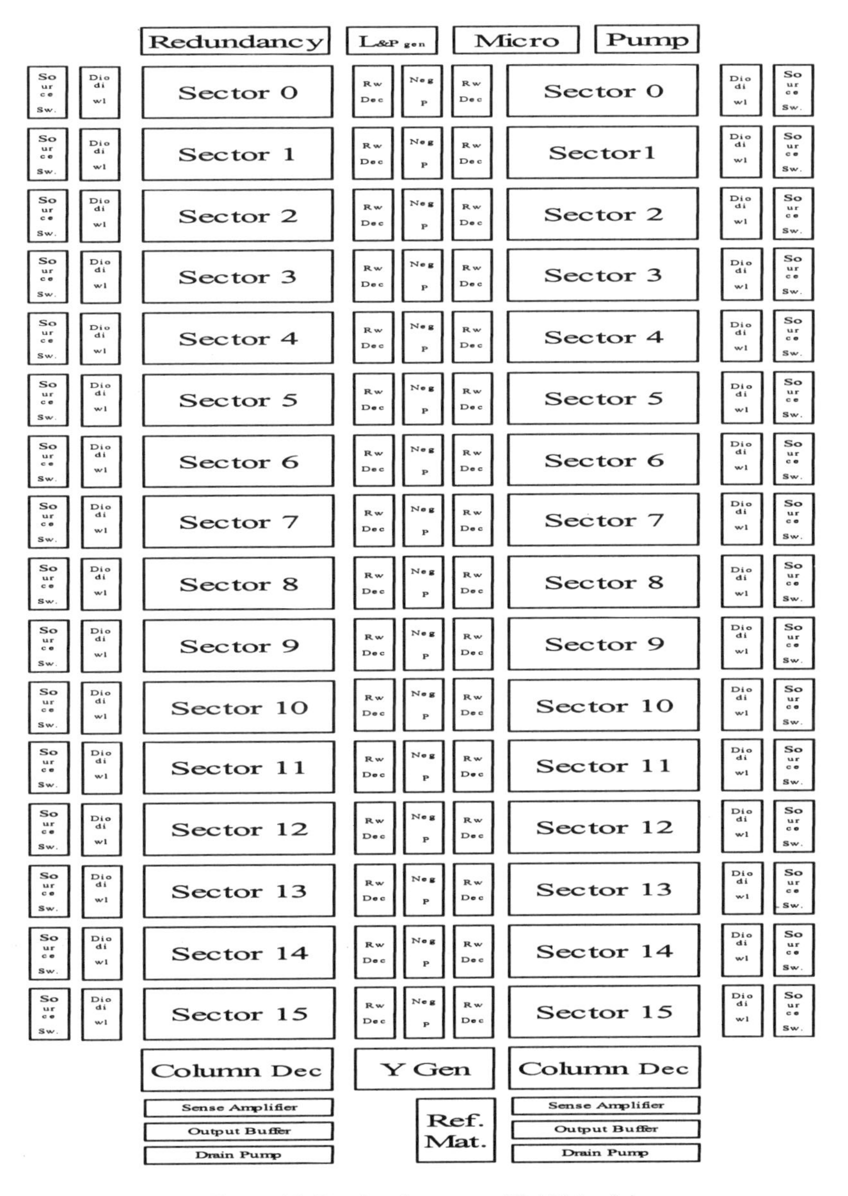

Figure 15: Device Structure (Half Matrix)

The same happens in the right half-matrix. As a consequence, in the decoding of the address of a word, a bit has the function to select the left or the right half matrix. Once the half-matrix is fixed the row can be selected. The partitioning in two half-matrix is due to the fact that the 16 Mbit memory is a doubled 8 Mbit memory (all the circuits are doubled except some of them), while the partitioning into sectors is due to the possibility of allowing the erasing of sectors beside the erasing of the whole matrix.

Within a half-matrix the partitioning in two halves of the sectors is due to the greater delay in the decoding of a row rather then in the decoding of a column. To reduce the delay it is convenient to divide the sector in two and double the final section of the decoding of a row for each half-sector. All these things are clearer looking at the Figure; this way can be seen that the row decoding is placed between the two half-sectors of a half-matrix.

Moreover in the Figure 15 can be seen that the charge pumps necessary to generate voltage boosted over the supply voltage to perform correctly the operations of writing (programming and erasing) are required.

It can also been observed the placement of the column decoding in the chip and subsequently the circuits that read memory cells (32 sense amplifier) followed by the output buffer. The link between these two is realised through a net to respectively give back as output the columns of the left and the right half-matrix.

Decoding

The decoding is the operation that allows the selection of the cells in the sector of the matrix, corresponding to the chosen address to perform the reading or the programming of such cells. Instead to erase it is enough to select the whole sector.

The cells of the Flash memory of a sector are organised in a matrix of MOS transistors with floating gate in which all drains of the cells belonging to the same column of the same sector are short-circuited to each other through a link called bit-line [19].

All gates of the cells belonging to the same row of the sector are linked through a link known as word-line. To conclude, the sources of each cell of the sector are driven by a source switch. This way, during the erasing, the switch will put the sources of the cells not to be erased to ground and turn to the required voltage the sources of the cells to be erased.

The operation of row and column decoding can be analysed more in detail. It will be assumed as a convention that the decoding is the totality of circuits that impose the drain or gate voltage to the cell, while the pre-decoding is considered to be the selection of row and column address.

Pre-decoding and decoding of a row

As previously said each half-matrix is divided in 16 sectors of 256 rows and 2048 columns (1024x2). So the whole matrix has a number of rows equal to 16x256=4096 (that is 2^{12}); therefore 12 bits are required to address without any uncertainty a word line and this is performed through the most significant bits of the A0-A19 address, that are those from A8-A19.

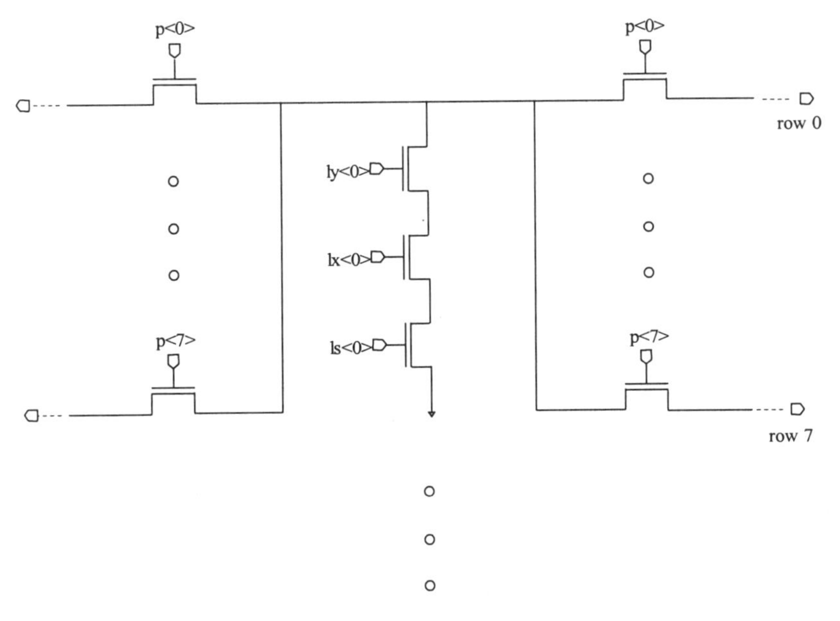

Figure 16: Pre-decoding of a row

In particular, the decoding of A16-A19 selects the corresponding sector (2^4=16) putting to the high value one of the 16 ls(0:15) signals. Once the corresponding sector is selected the chosen row must be selected. For each sector there are 256 rows, then, a list of three signals, p(0:7), ly(0:3), lx(0:7) are enough. These are obtained respectively with the address bit A8-A10 (2^3=8), A11-A12 (2^2=4), A13-A15 (2^3=8). As can be seen, there are 8x4x8=256 different signals on the whole.

Putting to the high value one of the p(0:7) means selecting one eighth of the 256 rows. Now putting to the high value one of the ly(0:3) signals means selecting a thirty-second of sector. To conclude, putting one of the lx(0:7) to the high value allows the selection of one of the 256 rows. This is what is called an “l” and “p” decoding whose architecture is realised through logical circuits implemented with NAND gates. The Figure 16 is an example of how the pre-decoding of a row can be implemented with a circuit.

Once the pre-decoding is finished the word line, to which the suitable voltage must be applied depending on the operation to be performed, is to be selected. The circuits necessary to apply such voltage have to take into account some features. In particular, in case of erasing, they must fix the gates of the cells belonging to the same row to a negative voltage internally generated starting from the supply voltage and regulated through suitable voltage regulators.

They must also be designed taking into consideration that the row of the matrix to which circuits are connected during the various operations presents a high capacitance load and delay due to high time constant. As a consequence, during any operations the gates of the cells of the selected row are in parallel offering an overall gate capacity of the order of some picofarad.

The interconnection of gate between a cell and the other of the same word line are realised in poly-silicon that has the capacitance load negligible, while it has a high resistance due to the lines length (of the order of KΩ). It can be guessed, so, that the RC associated to a line it is not definitely negligible.

As can be seen in the Figure 15 to the sides of the matrix there are diodes corresponding to each word line whose function is to link the line to the negative charge pumps which provides the voltage required during the erasing operation. Such voltage is not applied directly through the decoding circuits. The diodes are placed on both sides of the matrix because of a lack of space between a half-sector and the other one [20].

Pre-decoding and decoding of a column

The selection of the desired column is performed through switches realised with pass-transistors [15] [17]. Before examining how the pre-decoding takes place starting from the address, it will be described the link between drains of the cells belonging to the same bit-line. Such line consists of a link in *metal2* from which 4 connections in *metal1* depart for each sector of the matrix to which drains of the cells belonging to the same bit-line are linked.

Moreover, the *metal1* and the *metal2* are connected through pass-transistor. In particular, pass-transistors called *Y0* link the drains of the cells to the *metal2* through the *metal1*. Instead, pass-transistors called *YN* and *YM* link the *metal2* to the node called *yms*. Through it maintaining the voltage of 1V during the reading phase of a cell, the current to be compared to the reference one through sense amplifier, is drawn and a suitable voltage depending on the operation to be performed is applied.

The *yms* is kept to high impedance for unselected columns or during the erasing. It can now be seen more in detail how can be performed the pre-decoding, which is equivalent to apply a voltage to the signals *Y0*, *YN*, *YM* connected to the gate of the corresponding pass-transistors that result therefore selected.

The total number of columns of a sector of the half-matrix is 2048 but with regards to the decoding only 256 columns have to be addressed. In fact, dividing the columns of the sector into 8 groups of 256 columns each, all the columns are taken into account (256x8=2048). So, addressing one of the 256 addressable columns, the corresponding column of one of the eight parts in which the sector is supposed to be divided is selected.

In this way in output there are eight bits each coming from the same column of the eight groups. The 256 addresses are obtained through a simple combination of eight bits (2^8=256).

The eight bits correspond to the less significant bits of the A0-A19 address, which are those bits that go from A0-A7. In this way A0-A1 decoded serve to generate one of the four signals *Y0* (2^2=4), A2-A4 serve to generate one of the eight signals *YN* (2^3=8), A5-A7 decoded serve to generate the eight signals *YM* (2^3=8). As a whole, it is possible to select without any uncertainty 8x8x4=256 columns.

It is like supposing the 256 columns in eight groups of 32 columns each of that is then divided into eight other groups of eight columns. The Figure 17 shows what previously said.

As only one of the transistors corresponding to the signals *Y0*, *YN*, *YM* is switched on only one path is allowed contemporarily selecting the corresponding eight columns. In fact, the tree structure is repeated sixteen times. It must be underlined that during program operation the drain of the cells must be imposed to a voltage higher than the supply voltage. Such voltage obviously falls at the drain node of the selected transistor, so for making the pass-transistor working its gate voltage (that is the *Y0*, *YN*, *YM*) must be higher than the supply one too. It is, therefore, necessary to put particular care in the design of the circuits supplying the Y voltage.

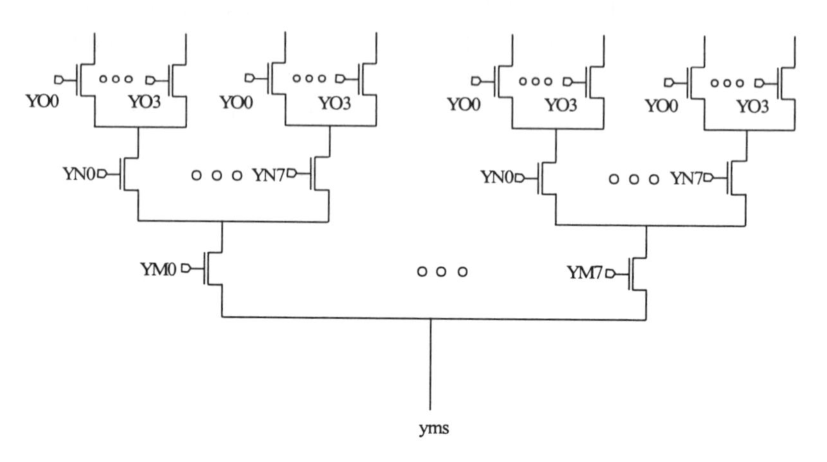

Figure 17: Pre-decoding of a column

To conclude it is important to underline that the pre-decoding of the column must take into account the redundancy that is the possibility to substitute a column containing a defective cell with an operating one.

The redundant columns are alternate to those of the matrix between the second and the third quarter of each half-sector. There are then four bit-line in *metal2* from which four connections in *metal1* are starting.

The first of the four *metal2* is referred to the first bit, the second one to the second bit and so on. In this way each output has at its disposal for each sector four redundancy columns. Thus the *YM* decoding of the faulty column is de-selected activating the *YR* decoding of the redundancy column while the *Y0* decoding is managed by the redundancy logic and not by the address on the input pins to choose one of the four possible paths in *metal1*.

Once the decision to redound one column is taken the columns having the same address and related to the other seven outputs will be redounded too. The eight bits of the address of the columns to be redounded is stored in the redundancy CAM (Content Addressable Memory), which are particular Flash memory cells.

Actually a CAM consists of five Flash cells, sharing the same floating gate, four of which are used for the reading and only one is used for the writing. The four cells also share the same drain. The advantage of this structure is due to the fact that it is possible to simultaneously program the five cells.

The choice to make a column redundant instead of a row is due to the fact that there are few probability of a short-circuit between two adjacent poly-silicon word line and that 12 address bits corresponding to the row to be redundant should be used instead of the eight required by a column.

The Read Path

After having examined the device structure and how to perform the addressing of a word it is now possible to take into account how to execute the reading of a memory cell. It must be remembered that the read path implies an input stage, which guarantees a certain noise immunity that could cause undesired transitions.

This problem must not be undervalued as a new reading cycle starts at each transition in the input, then a spurious transition due for instance to the noise on the supply or on the ground, would result in the interruption of the reading cycle to start a new one compromising the operation and the access time if the addressed byte was always the same.

Moreover a fundamental characteristic which the input stage must satisfy is the interfacing with the logic family differing from the CMOS. In this way the logic threshold of the input buffer must be referred to the TTL levels too. As it was said in the previous chapter the read operation is divided into various phase each of which has a fixed timing so that all the phases will be synchronised through pulses of clock of suitable period rising from a starting pulse called ATD (Address Transition Detection) which must be generated each time a level transition takes place on one of the addresses pins. So, for each new reading an ATD generator creates such pulse, which allows the synchronisation of all the reading phases starting from the pulse itself. Along the read path, immediately after, there is the sense amplifier which allows the reading of the information of the addressed cells and the output buffer.

The latter has the function to provide the output of a signal varying between the logic levels chosen within reasonable time. As the stage must drive high load capacitance the output buffer must be able to provide high output current. To this reason is related the problem of the transition noise. In fact, the supply and the ground lines are subjected to the fluctuation of the potential due to the voltage drop on the parasitic resistive and inductive elements that take place on the connections and on the bonding wires (these links the pad on the integrated chip with the package pins and can be represented by an inductance of the order of some nH). The voltage drop on the rows is the greater the higher is the injected or absorbed current. So, on one hand a higher current is required to drive considerable load, on the other hand the higher this current is the greater are the fluctuations of voltage due to the parasitic effects on the line.

The matter is more relevant the higher is the number of outputs simultaneously switching in the same direction (0→1 or 1→0) as the supply or ground fluctuations are proportional to this number. The principal consequence is that the circuits suffering mainly of such problem are the output stage and the sense amplifier that clearly shows the need for a careful input stage design.

After this overview on the read path it is possible to examine more in detail the sense amplifier.

Sense Amplifier

The sense amplifier is the circuit that allows the reading of the information stored into a memory cell.

The principle of sensing is based on the comparison through a buffer between two voltages, one related to the cell to be read and the other related to a reference cell. As a consequence of this, the buffer gives back at the output a value, which can be high or low depending on the result of the comparison. To do so it is necessary:

- to convert the current of the cells (both the reference and the matrix one) into a voltage available at the input differential stage of the buffer;
- to avoid that the drain voltage in the reference and in the matrix cells exceed a value near to 1.5 V to prevent a problem called soft-writing;
- to de-couple the low capacity node of the input differential stage of the buffer from the high capacity node due to the bit-line and to the column decoding;
- to speed-up the read equalising the addressed bit-line with the reference side (to speed up the comparison it is necessary that the unbalancing of the voltages at the comparison node takes place starting from the same levels);
- to provide a first amplifying stage.

Afterwards the conversion of the output analog signal of the sense amplifier into a digital one, which can drive the output buffer is to be done. In the Figure 18 it is shown a possible implementation of the sense amplifier called differential architecture [16] [10].

In this kind of architecture the information of the matrix cell is compared with a reference cell. This allows to minimise the variations of process technology, of temperature and of supply treating all of them as contributes of common mode. In fact, it is known that, as the differential stage has a high CMRR, it rejects noise of a common mode. Moreover such configuration allows to obtain a reference current from a virgin cell.

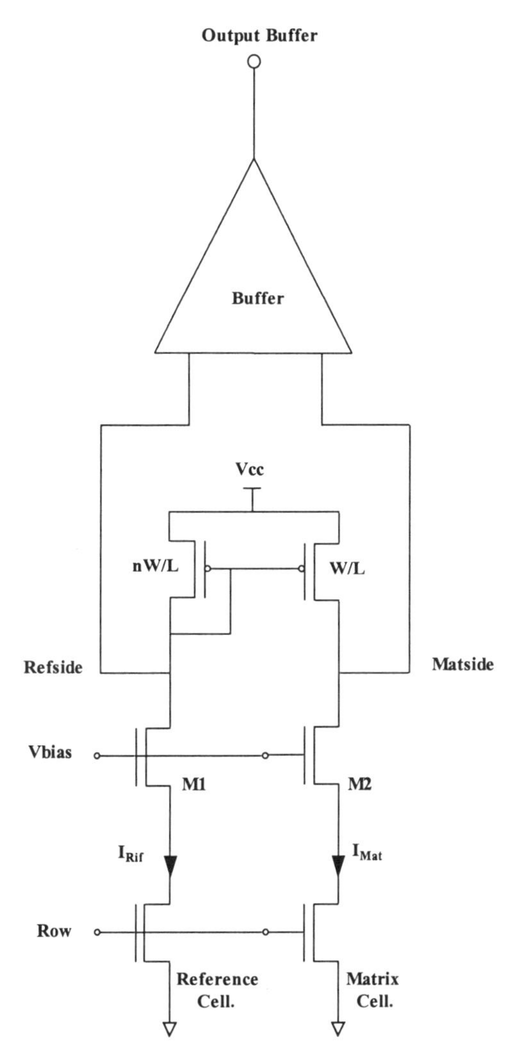

Figure 18: Differential architecture of the sense amplifier

As can be seen in the Figure 18 the matrix cell current and the reference one are compared through a differential stage by a conversion I-V through the p-mos transistor current mirror. Such transistors have the characteristic of featuring shape factors one multiple of the other. The transistor connected to diode on the reference side has its "W/L" of n times the one of the p-mos transistor on the matrix side. It is, then, quite easy to understand the behaviour of the circuit. In fact, the reference current imposed by the reference cell is mirrored into the matrix side with 1/n ratio. On the other hand, the matrix cell, having a drain and a gate voltage value depending on whether the cell is programmed or erased, imposes its own current on the same side.

As a consequence the voltage at the *matside* node, to which a parasitic capacitance load is associated, will rise or fall depending on whether the matrix cell conducts a current higher or lower than I_{ref}/n.

The current by which the node charges or discharges has an absolute value of $|I_{ref}/n—I_{Mat}|$. When the cell is programmed, then, $I_{Mat}< I_{ref}/n$ and the output node is set to V_{cc}. When the cell is erased or virgin $I_{Mat} >I_{ref}/n$ and the output node is set to ground.

The presence of the cascoded structure, M1 and M2 transistors, is of huge importance. In this way $V_{BL} = V_{BIAS} - V_T - \sqrt{2I/\beta}$. For a native transistor, with a low doped channel, the β factor is quite high, so that $V_{BL} = V_{BIAS} - V_T$. Then if the voltage of the bit-line tried to exceed the limit imposed by such relation the M1 and M2 transistors would switch off, preventing, in this way, the bit-line voltage from exceeding a fixed value (1V).

Moreover the cascode has another function: to de-couple the high capacitance of the bit line-line (pF) from the output node (fF) allowing a faster charging of such node and avoiding spurious writing phenomena (soft writing) during the reading phase. If the drain voltage is not kept low, a high number of drain cycles can cause the undesired writing of the cell.

The parasitic capacitance of the bit-line is fundamentally due to three factors, that is, the parasitic capacitance towards the bulk which derives from the parallel of the capacitance of the depletion region of the drain junction of all the Flash cells belonging to the same column, the capacities related to the metal wires which link all the cells of the column, and the parasitic capacitance associated with the column decoding transistors.

Program Operation

The device is programmed for byte; this operation can be done by mean a sequence of instructions on the input pins of the chip. Embedded Program Algorithm will produce automatically the programming pulses of a suitable length, verifying also that Flash cell has the appropriate threshold voltage. Typically this operation is carried out in some tens of microsecond (15 ÷ 25 μsec).

Often programming and writing are confused, generally "write" is referred to the set of program and erasing instructions.

Instructions definition

Through the input pins the list of instructions is sent to the CUI (Command User Interface), which consist of a state machine. Such unit has the function of reading and storing in suitable registers the instruction set, it also memories the data required performing the instruction.

Command	Length	Bus Write Operations—16 bit mode							
		1st		2nd		3rd		4th	
		Addr	Data	Addr	Data	Addr	Data	Addr	Data
Read	1	X	F0						
Reset	3	555	AA	2AA	55	X	F0		
Program	4	555	AA	2AA	55	555	A0	PA	PD
		Bus Write Operations—8 bit mode							
Read	1	X	F0						
Reset	3	AAA	AA	555	55	X	F0		
Program	4	AAA	AA	555	55	AAA	A0	PA	PD

Note: X Don't Care, PA Program Address, PD Program Data

Table 14: Instructions table

Instructions are sent to the CUI within several bus cycles forcing the three control pins "*OENPAD*" (*output enable*, active low), "*CENPAD*" (*chip enable*, active low) and "*WENPAD*" (*write enable*, active low).

The programming operation is selected by writing fixed list of addresses and data into the CUI, as they are specified in the Table 14. To send any instruction it is necessary to force *WENPAD* and *CENPAD* to low logic level while *OENPAD* must be brought to a high logic level; the register of addresses latches on the last falling edge between *WENPAD* and *CENPAD*, while the register of data latches on the first rising edge of *CENPAD* or *WENPAD* signal.

Programming

Through the program operation it is possible to write "0" in a specified location, on the other hand through the erase operation a "0" is converted into "1". Programming can be performed by 8 bits (byte) or by 16 bits (word), and it requires four cycles.

The first two are unlock cycles, and they have the function of preventing undesired programming operation. The third cycle is a setup instruction for programming, finally, the fourth is the data writing cycle. The addresses register is written on the falling edge of *CENPAD* or *WENPAD* (the one which descend last) while the data register on the rising edge of the first signal between *CENPAD* and *WENPAD*.

The programming operation starts exactly on the rising edge of the first signal between *CENPAD* and *WENPAD*. After performing the list of instructions to access the inner programming algorithm, the system does not have to do any other check, as the device will automatically provide suitable programming pulses internally generated and will verify that the threshold voltage of the programmed cells is correct.

When the process is completed, the device go back to the read modality and the data, related to the addressed byte at that moment, will be reported to the output.

To program groups of 8 bits the input pin "*BYTENPAD*" must be set to a high logic level, at the opposite, if such pin is at a low logic level the program operation takes place by groups of 16 bits.

The CUI

The Command User Interface (CUI) is realised through a finite state machine (FSM), which recognises the arrival of a correct instruction list, updating the corresponding control flag.

The state machine can choose between alternative state sequences and belong to the class 4, which means both the transition and output states are conditioned [8] [9].

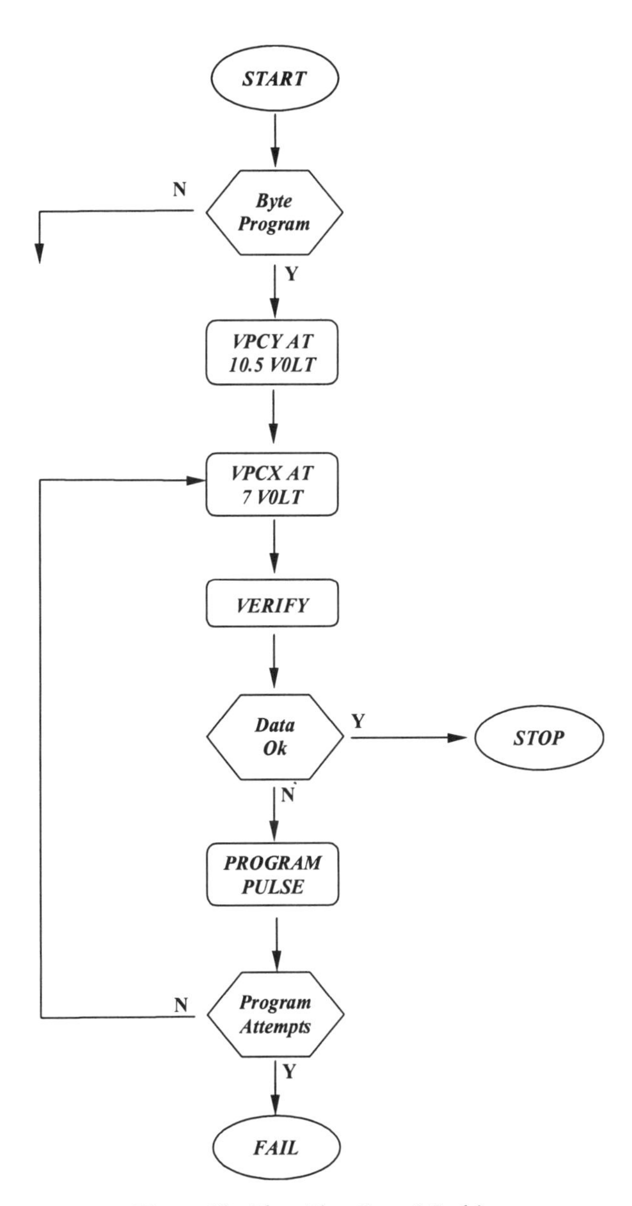

Figure 19: Algorithm State Machine

The Algorithm State Machine (ASM) represented in Figure 19 through a flow chart describes the program operation. The *VPCX* signal is the voltage applied to the gate of the Flash cell, while the *VPCY* signal is the voltage applied to the pass transistor gates of the column decoding.

The synthesis of the machine is automatically realised through software stored in a ROM. The check of a certain configuration to the address and data pins takes place through simple logic gates, NOR and NAND. A signal indicating which operation is being performed is then generated. Such signal is memorised in a suitable register till the end of the operation.

The internal microprocessor

As can be guessed the Flash memory is an embedded system with microprocessor, thus consisting of both a hardware and a software part.

One of the main advantages of this system lies in its versatility, such structure actually allows a complete reprogramming through software without radically changing of the layout, and on the other hand it requires only the change of few masks related to the ROM in the process flow.

The architecture of the microprocessor is quite simple, consisting of a *program counter*, a *ROM*, a *logical section*, an *instruction decoder* and an *output buffer*. A clock working at 25 MHz scans the microprocessor, it also features eight instructions and manages twenty outputs each of which manage various sub-circuits of the device.

In Figure 20 the block scheme of the processor is shown and it is possible to see the *ROM*, addressed by the *program counter*, and containing the software.

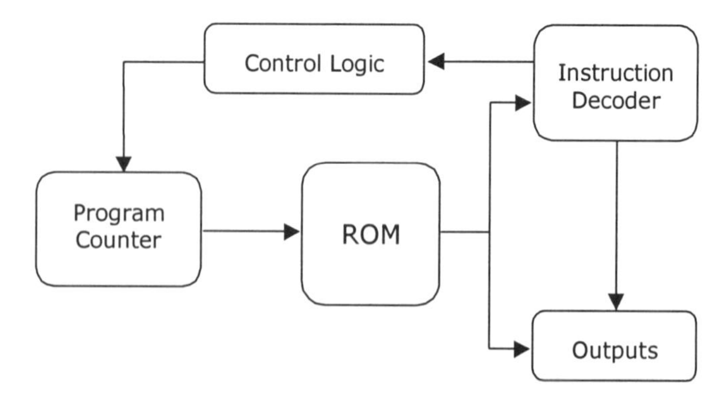

Figure 20: Block scheme of the microprocessor

The *ROM* is a 112 x 12 bit memory organised in 28 rows and 48 columns matrix. Its outputs are 12, three of which form the instruction code and the remaining are used as argument of the instruction itself. Through a ratio logic, the *ROM* can be read, and the data synchronised with clock is stored into a register and sent to the output. The *ROM* is addressed with seven bits, by the *program counter* which allows to proceed sequentially while executing the program, to block the scanning until an external event occurs or to jump over a fixed line of the program. The different possibilities to be done obviously depend on the executing instruction.

The *program counter* consists of a synchronous counter, which manage through seven output bits the three options previously described.

The first three output bits of the *ROM* allow to implement eight different instructions each of which is executed with a clock and checked through simple three inputs *AND*. Jump instruction are three, one unconditioned (*JMP*) and two conditioned (*JIF* and *JFN*). Two instructions are of comparison (*CMP* and *CMP2*), one of waiting (*ALT*) and two of activation of the various outputs (*STO* and *STO2*). Actually, two instructions are redundant (*STO2* and *CMP2*) so it is correct to talk about six instructions.

The three jump instructions are with absolute addressing; the operator, which indicate the address which jump to, consist of the seven less significant bits of the *ROM*. The *JMP* effect is the forced loading of an address. *JIF* (jump if) and *JFN* (jump if not) are conditional jump instructions. Each time a comparison instruction is performed (*CMP* or *CMP2*) a flag is set to "1" or "0" and memorised depending on the result of the comparison. The *JIF* instruction is performed when the flag is "1" otherwise it is ignored, similarly *JFN* is performed only if the flag is "0". The *ALT* instruction wait that the signal associated to its operator get to a high logic level stopping the program counter, such instruction is useful implementing delay cycles of the program. The *STO* and *STO2* instructions enable the microprocessor outputs by setting them to the same value of the operator in the *ROM*.

The program algorithm

The inner algorithm automatically manage previous operations, without requiring any external event, moreover information on the running algorithm and eventually occurred errors are simultaneously given on the external pins in order to inform the user about the state of the state machine. Internal algorithm is performed and managed by the small synchronous microprocessor previously described.

The program algorithm is quite simple and allows the writing only of a "0". First of all it compares the external byte representing an address with the one currently present in memory, if they are equals the algorithm is ended, otherwise a program pulse will be required. Such pulse will be sent only to the bits to be programmed and not to the programmed bits at the specified address.

The end of the program pulse is set counting a certain number of clock cycles. At the end the algorithm checks that the stored data is correct. If the result is positive the algorithm is ended, otherwise a new program pulse will be generated, then a new data check will be performed until the maximum number (~ 400) of program pulses is exceeded which indicates a fault Flash cell.

```
0    CMP   BWPROG
1    JIF   25
...
25   ALT   VPCOK
26   do nothing
27   do nothing
28   do nothing
29   do nothing
30   do nothing
31   do nothing
32   CMP   BWPROG
33   JIF   56
...
56   ALT   VPCOK
57   STO2  VERIFY
58   CMP2  DATOOK
59   JIF   67
60   STO   MPROG
61   ALT   ENDPULSE
62   CMP2  SUSPREQ
63   JIF   117
64   STO   INCTENT
65   CMP2  MAXTENT
66   JFN   56
67   CMP   BWPROG
68   JIF   114
...
114  STO2  SWXATVCC
115  ALT   NORMOP
116  STO2  CUIRES
117  STO2  STOREADD
118  STO2  SWXATVCC
119  ALT   NORMOP
```

Figure 21: The program algorithm

The Figure 21 shows in detail a part of the algorithm related to programming operation.

The *BWPROG* signal at high logic level indicates the starting of the programming operation. *VPCOK* reports whether the value of the voltage applied to the gate of the Flash cell is correct or not.

The *VERIFY* signal indicates, at a high logic level, that a check operation is being processed (*program verify*, *erase verify* or *depletion verify*). *MPROG* is the program pulse and *DATOOK* report whether all the data bits have been correctly programmed. *ENDPULSE* fix the end of the program pulse.

The *SUSPREQ* signal at a high logic level suspends the programming. Usually, it is due to external conditions given by the user. The *SXWATVCC* signal switches the gate voltage of the Flash cells and of the pass transistors to the reading mode voltage. *NORMOP* report whether the operation has been correctly processed or not. *STOREADD* activates the memorising of the current address in case of interruption of the operation. *CUIRES* resets the CUI and fix the end of an operation, resetting the device in reading mode.

Device simulation

Particular care has been put on the correct execution of the internal program algorithm. Simulations have been performed without the matrix of the Flash memory cell to avoid the overload of the calculators. In such conditions a simulation regarding the programming process lasts 6-8 hours.

The Figures 22 and 23, obtained by the simulation, report the signal allowing the functional check of the device in program modality. From the simulation, in addition to the signals of the programming algorithm, it is possible to observe both the input signals (stimuli) and the other signals, which allow an easy functional check of the device.

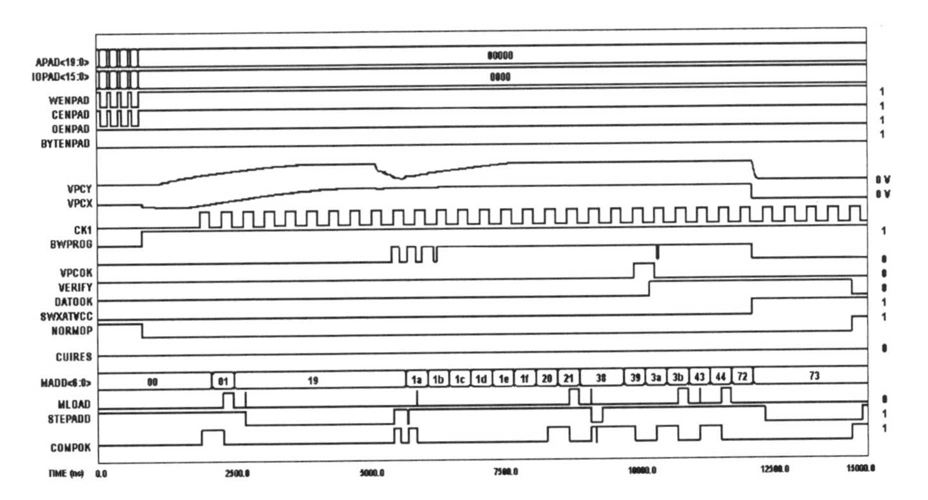

Figure 22: Device simulation

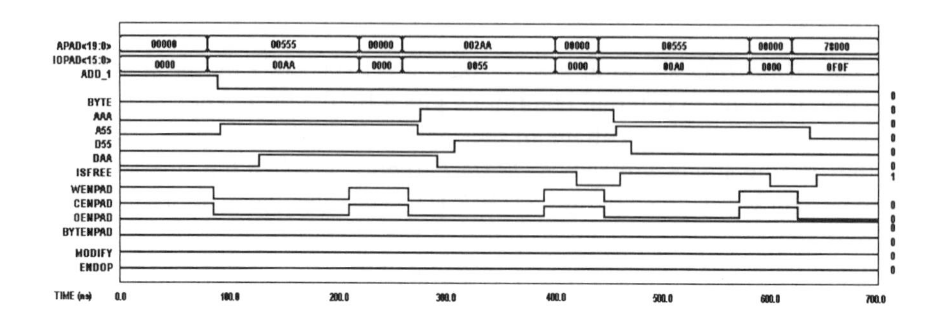

Figure 23: Device simulation

The *APAD<19:0>* and *IOPAD<15:0>* bus, the signal *WENPAD*, *OENPAD*, *BYTENPAD* and *CENPAD* are the inputs of the device. The *IOPAD<15:0>* bus is also the device output.

Through the address bus *APAD<19:0>* the user can define the instructions to be given to the CUI to execute the programming operation, in this case the instructions are given within four bus cycles.

The data bus *IOPAD<15:0>* is the output of the device, but it also represent the data and control output for the various operations performed by device. *CENPAD* can enable or disable the device, *WENPAD* enables the cells writing and *OENPAD* enable the output bus of the device.

Two analog signals, *VPCX* and *VPCY*, are respectively the voltage applied to the gate of a Flash cell and the voltage applied to the gate of a column decoding transistor, and they change in accordance with the operation to be performed. The logical signal CK1 is the clock, which scans the operations of the microprocessor.

The *MADD<6:0>* bus reports the output of the address counter of the microprocessor and it indicates the memory location of the ROM where the next reading will be done (it corresponds to the algorithm line related to the operation in the hexadecimal system).

The *MLOAD* signal indicates to the counter the contemporary loading of the *MADD<6:0>* signal, it is "on" when a jump instruction (*JMP*, *JIF* or *JFN*) is satisfied. *STEPADD* is always high except during the *ALT* instruction. Finally *COMPOK* is at high logic level if the operator of a comparison instruction is also at high logic level.

The logic signals *AAA*, *A55*, *DAA* and *D55* are check signals; in the first three bus cycles of programming, they report whether the data or the address has been decoded. ISFREE disables the programming and the erasing of the protected sectors.

The device simulations show the functional check of the device in programming mode.

Erase Operation

The behaviour of the device during the erase operation will be explained, starting from the state code needed to turn the CUI on to the operations run by the state machine.

Starting of the CUI state machine

One of the operations available within the device is the erase. On the analysed Flash memory, it is possible to perform two erasing operations: sector erase, which involves the erasing of only some sector of the whole memory matrix, chip erase, which allows the erasing of all the cell of the device.

To perform both these operations first all, it is necessary to send a list of fixed state code, both to the address pins and to the data pins. In this case the erase sequences are reported in Table 15 and 16. Such instructions list, sent to the CUI, make the corresponding control flag to be turned on.

	Bus Write Operation											
Command	1st		2nd		3rd		4th		5th		6th	
	Addr	Data	Addr	Data	Addr	Data	Addr	Data	Addr	Data	Addr	Data
Chip Erase	555	AA	2AA	55	555	80	555	AA	2AA	55	555	10
Block Erase	555	AA	2AA	55	555	80	555	AA	2AA	55	BA	30

Table 15: Instructions in 16-bit modality, Byte = V_{IL}

	Bus Write Operation											
Command	1st		2nd		3rd		4th		5th		6th	
	Addr	Data	Addr	Data	Addr	Data	Addr	Data	Addr	Data	Addr	Data
Chip Erase	AAA	AA	555	55	AAA	80	AAA	AA	555	55	AAA	10
Block Erase	AAA	AA	555	55	AAA	80	AAA	AA	555	55	BA	30

Table 16: Instructions in 8-bit modality, Byte = V_{IH}

The algorithm, ASM (Algorithm State Machine) chart, which describes the process, is reported in Figure 24.

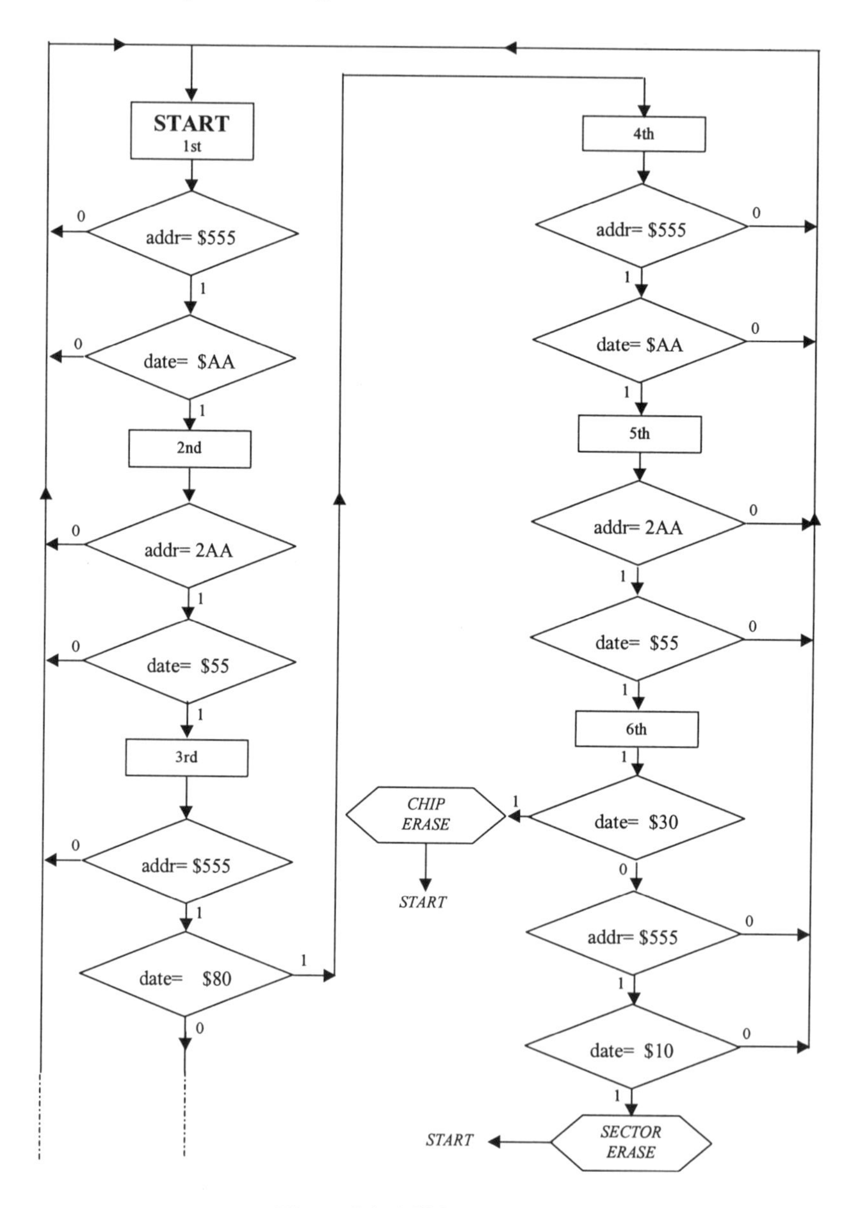

Figure 24: ASM Erase chart

The choice of this one as the suitable state code is due to the fact that it seems to be the most efficient from the point of view of the circuit realisation. A signal that indicates which operation is being performed is memorised in a suitable register, till the end of the operation itself.

The erase algorithm

Once the right state code sequence is finished, the microprocessor, which runs the erase algorithm, is started. Such algorithm is really complex as it must guarantee that all cells are programmed before erasing them, avoiding to make them depleted. Once the program pulse is given it must verify that the threshold of the erased cells is low enough (erase verify) but not too much (depletion verify) providing respectively more program pulses or soft programming pulses.

The erase algorithm starts scanning the 32 sectors till the first to be erased is revealed. Once such sector is localised, the algorithm programs all and only its bits that result to be 1, with the same modality of the *byte program algorithm*. Successively, if necessary, the operation is repeated for the other sectors. This phase is called *all0*, as all the cells are set to "0". At this point, the electrical erasing starts, giving the program pulse simultaneously to all sectors concerned so that the overall time of the algorithm is minimised. At the end of the pulse, the correct erasing of all the cells of the sectors involved in the erase is checked, setting some flags that indicate which sectors require another erase pulse; the voltages necessary for the erasing will be given only to them.

To reduce the overall period of the erase operation, for each sector, the address of the first cell with faulty erasing encountered during the scanning of the sector itself is memorised in a small internal RAM, after every pulse. The following check will start from this address. Erase pulses are given till the maximum number allowed (~ 300) is exceeded and in this case an error flag is lighted, or all sectors are correctly erased and in this case the algorithm proceed to the next step, the *depletion verify*.

The *depletion verify* operation is used to locate the cells with a voltage threshold too low (even negative), which might conduce current even if their gate is grounded. Their current, even if they were not selected, would sum to the current of the selected cell belonging to the same bit line of the same sector, causing a faulty reading. The threshold of the depleted cells is increased through the soft programming process till it reaches an acceptable value.

During this step a gate voltage lower then the voltage used for normal programming is given to the cell (the drain voltage is the same).

As it is impossible to univocally locate the cells that could be depleted, and to avoid to soft program too heavily the cells during this phase, various voltage steps applied to the gate have been provided. So, at the beginning, the first column of a scanned sector is checked to sense if the current is higher than it should be. If this happens, a soft programming pulse is given starting from the higher step. After that the same column is scanned again in case a cell is still depleted. If this happens another soft programming pulse is given to the cell, but always corresponding to a lower step of voltage. Once the scanning of the column is completed, the column is verified, and a new soft programming pulse is given to all these cells which might need it, this time corresponding to the next step of voltage.

This process guarantees a strong programming of the depleted cells and a weak programming of the normal cells, whose voltage will result almost unchanged.

The sector/chip erase algorithm, described above, which is run by the state machine, is represented in Figure 25.

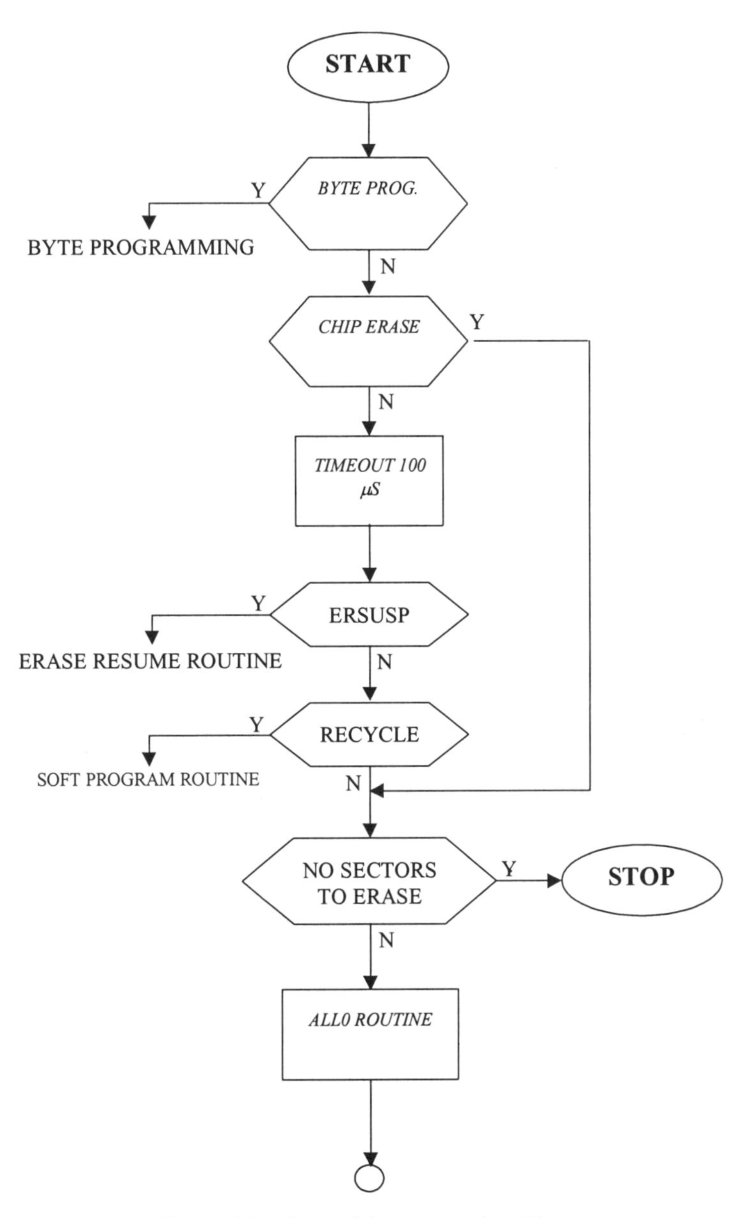

Figure 25a: Sector/chip erase algorithm

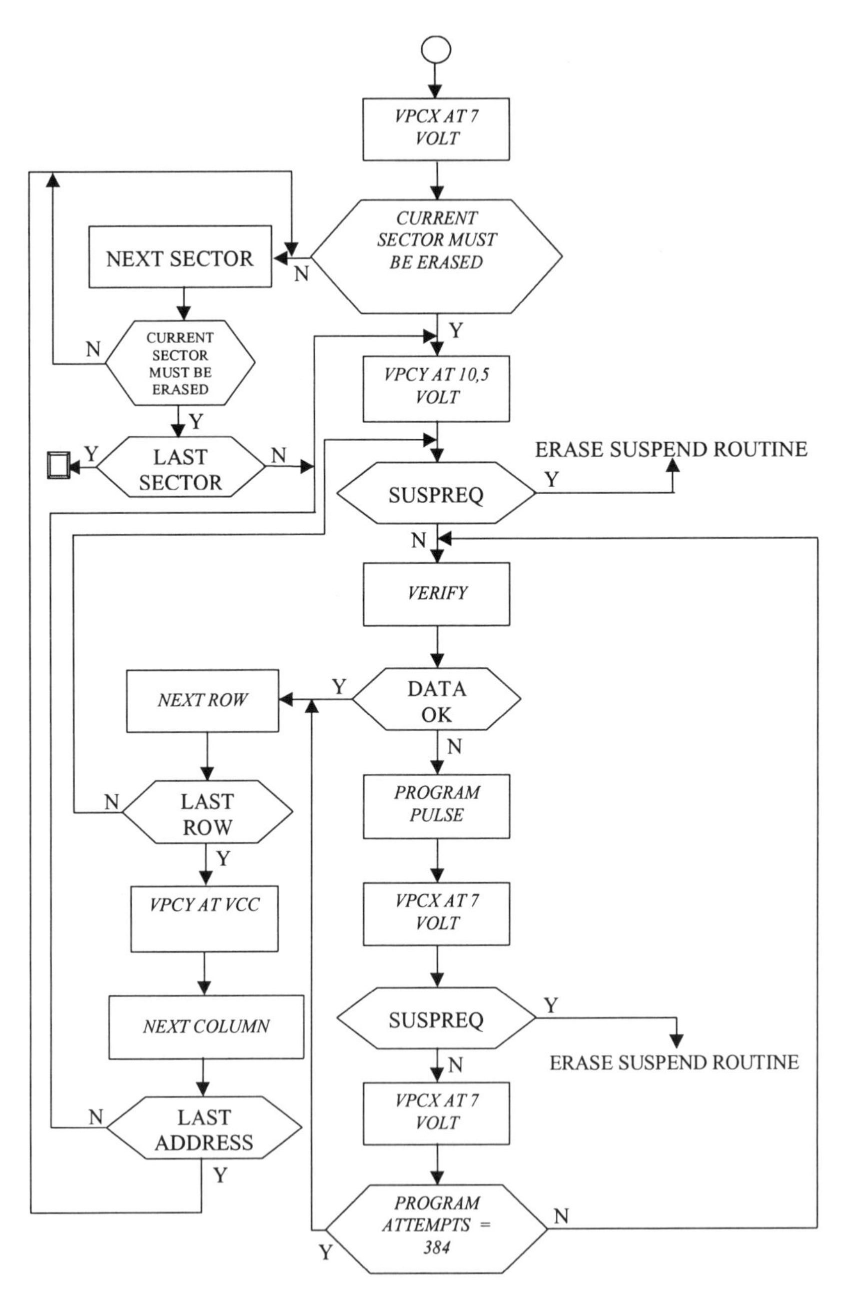

Figure 25b: Sector/chip erase algorithm

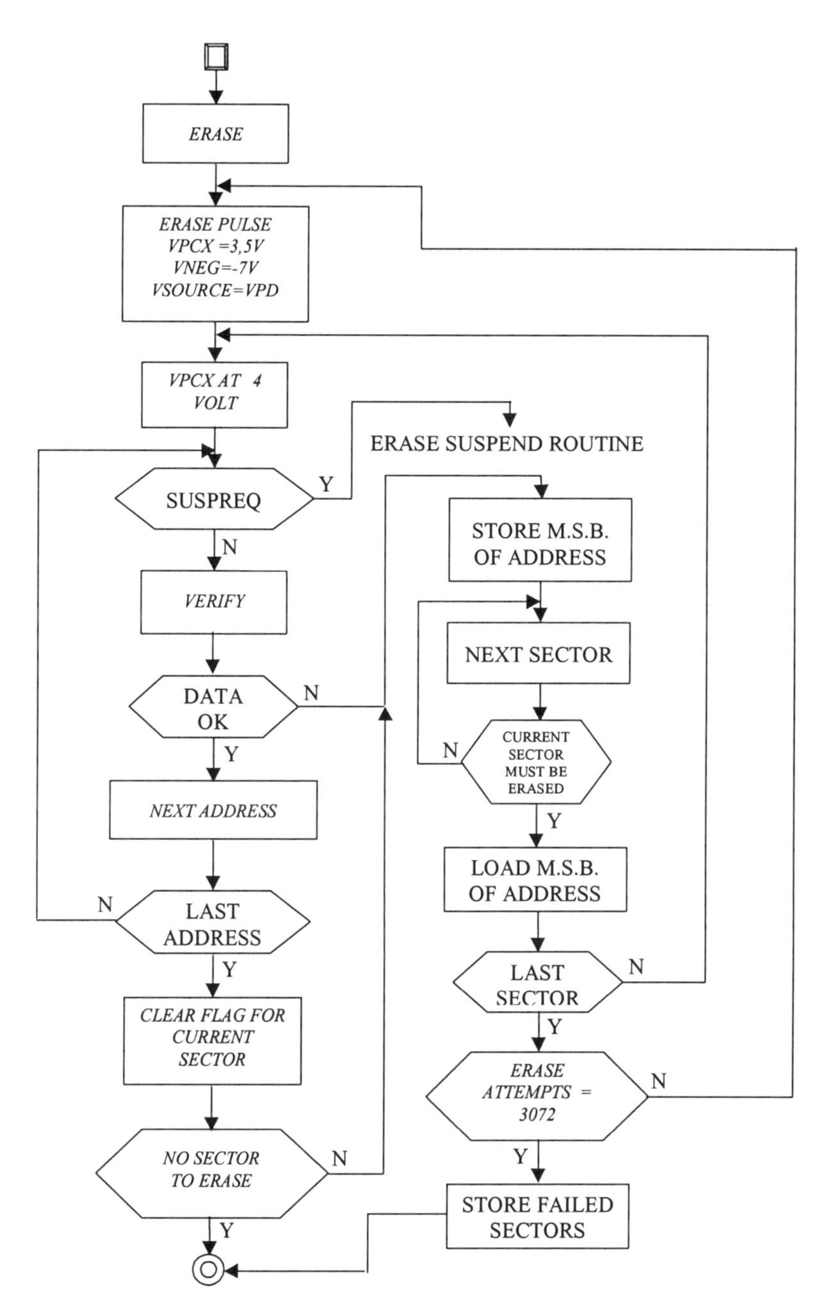

Figure 25c: Sector/chip erase algorithm

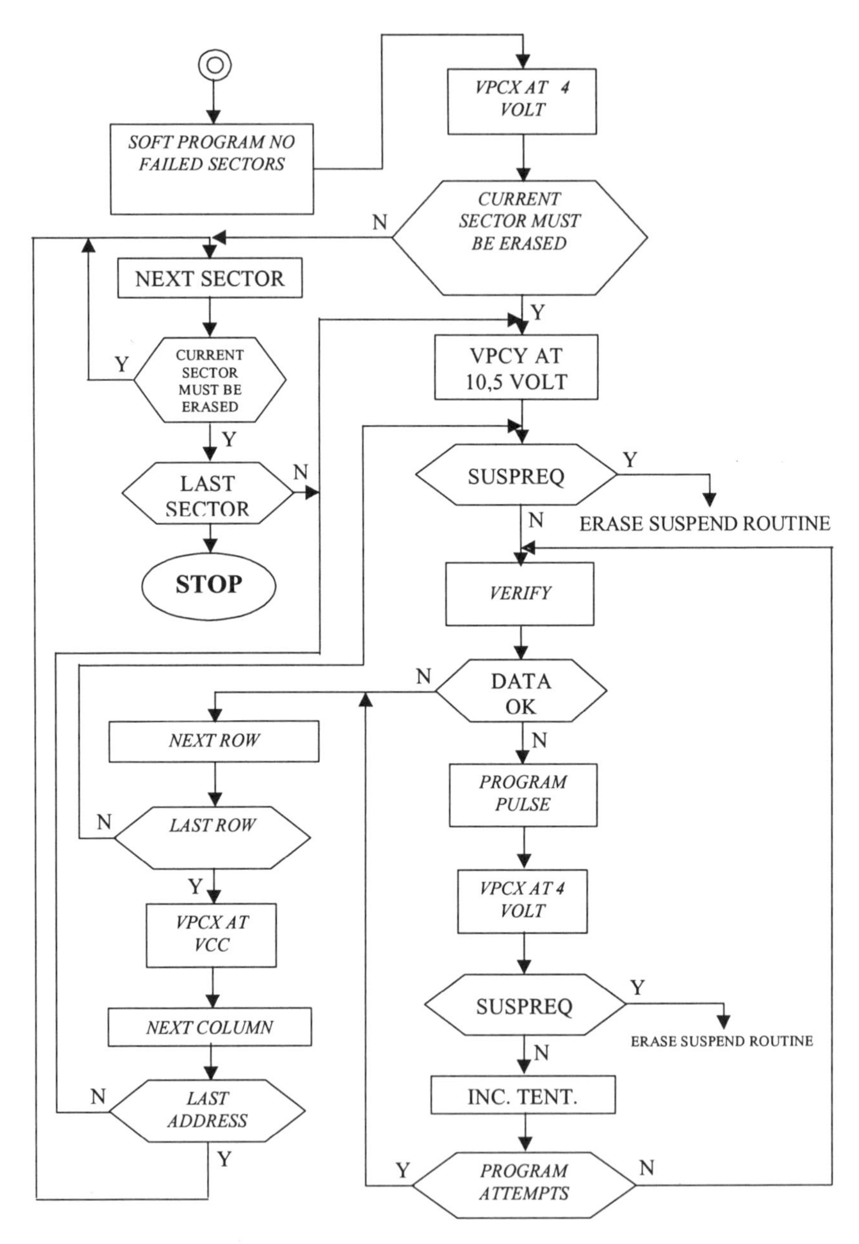

Figure 25d: Sector/chip erase algorithm

Conclusions

Our work examines a low voltage single supply 16 Mbit Flash memory and better known into the market with the trade mark name of M29W160BT.

First of all it is described the elementary cell of memory that properly stores information (bit). We tried to point out technological characteristics of the cell and physical processes involved by programming and erasing operations.

Successively we explain step by step full chip simulations. During the design of a chip simulation tools allow to test its right work. Given some models offered by technology and used to represent devices found into integrated circuits, simulators allow to solve complex equations of a circuit. It is also possible to compare simulations given by different simulators or by the same simulator on the same device but in different conditions. The features of these tools have been described.

Electrical and functional characteristics of the device (datasheet) have been examined.

The reading operation, the circuits involved and the deriving problems are shown.

After that we focus our attention on programming operations themselves. The results of the simulations show the signals involved and the control given to the finite state machine step by step for the right decoding of the operation.

As a consequence we speak about the erasing operations and particularly about the complex internal algorithms processed by the FSM to carry out the operation.

The wide explanation shows how complex is the structure of a Flash memory. In fact, we speak about "system on chip" because the device bear inside all together a matrix of cells for the storage of information, a FSM able to run the execution of the different operations and their relative controls, a circuitry creating positive and negative voltages necessary to program and erase, and finally all those circuits devoted to the read of the cell memories. It is also clear the complexity of the flow scheduled to carry out the simulation.

References

[1] P. Cappelletti, C. Golla, P. Olivo, E. Zanoni, *Flash Memories*, ed. Kluwer Accademic Publishers.

[2] P. Pavan, R. Bez, P. Olivo, E. Zanoni, *Flash memory cell—An overview*, Proc. of the IEEE, 85.

[3] D. C. Gunteman, I. H. Rimawi, T. L. Chiu, R. D. Halvorson, D.J. McElroy, *An electrically alterable nonvolatile memory cell using a floating gate structure*, IEEE Trans. on Electron Devices, 26.

[4] G. Verma, N. Mielke, *Reliability performance of ETOX based Flash memories*, Proc. IRPS.

[5] D. Kahng, S. M. Sze, *A floating gate and its application to memory devices*, Bell Syst. Tech. J., 46.

[6] R. Bez, *Flash Memory*, SGS-Thomson.

[7] S. T. Wang, *On the I-V characteristics of floating gate MOS transistors*, IEEE Trans. on Electron Devices, 26.

[8] Comer, *Digital and State Machine Design*, ed. Oxford Press.

[9] C. Clare, *Designing Logic Systems using State Machine*, Mc Graw Hill Book Company.

[10] M. Annaratone, *Digital CMOS Circuit Design*, ed. Kluwer Accademic Publishers.

[11] *Eldo User's Manual*, Document Number 310101 for Software Version V5.7_3.1 November 1998 Revision 7.3, Mentor Graphics.

[12] *OpenBook Cadence*, Release 9502.

[13] *Device Datasheet M29W160BT*, (M29W160BB), 16Mbit (2Mb x 8 or 1Mbit x 16, Boot Block) Low Voltage Single Supply Flash Memory.

[14] *PowerMill Reference Guide*, Release 5.3, May 1999, Synopsys.

[15] B. Ricco', F. Fantini, P. Brambilla, *Introduzione ai circuiti integrati digitali*, ed. Zanichelli Telettra.

[16] Randall L. Geiger, Phillip E. Allen, Noel R. Strader, *VLSI Design Techniques for Analog and Digital Circuits* ed. Mc Graw Hill International Editions.

[17] Jan M. Rabaey, *Digital Integrated Circuits, a design perspective*, ed. Prentice-Hall International Inc.

[18] R. Muller, T. Kamins, *Device Electronics for Integrated Circuits*, John Wiley & Sons.

[19] C. Hu, *Nonvolatile Semiconductor Memories, Technologies, Design, and Applications*, IEEE Press.

[20] B. Prince, *Semiconductor Memories, a Handbook of Design, Manufacture and Applications*, John Wiley & Sons.

VHDL Design, DFT, ATPG & Layout Implementation Service of a Digital Block for a DAC Converter

A. De Capoa, P. Lamanna, S. Piccioni

Accent S.r.l.

Via F. Gorgone 6, 95030 Catania - Italy

E-mail: (andrea.decapoa, pasquale.lamanna, stefano piccioni)@accent.it

Introduction

This work consists in re-designing a digital block of a DAC (Digital Analog Converter) with 20 bit input, 2 stereo channel (Left, Right) and 20 MHz clock for a car-audio application. Internal Accent designers have already designed this block. We were asked to improve the architecture in order to make the chip smaller (and cheapest!).

In the Figure 1 we can see the original layout of the whole DAC chip project:

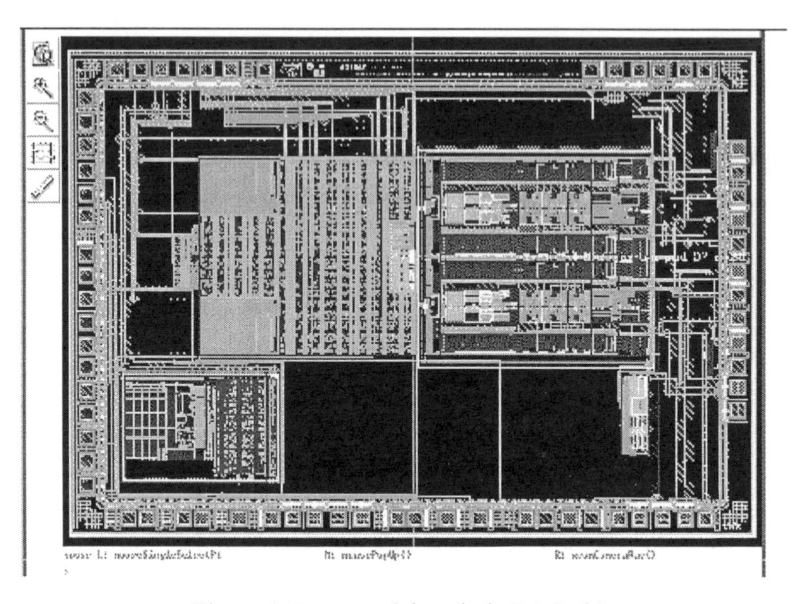

Figure 1: Layout of the whole DAC chip

Below there is the portion of the previous layout representing the digital part under examination. From now on we will refer to this digital block as **DB-DAC** (Digital Block of DAC):

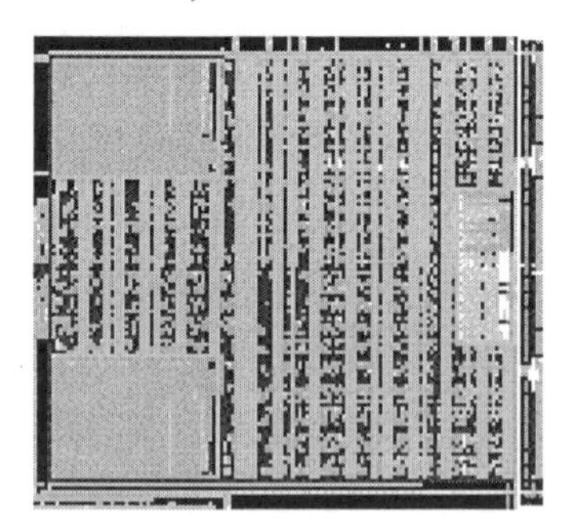

Figure 2: Layout of the digital block of DAC chip

The project flow was divided in three major skill areas in order to better fit with our background competencies. Thus each of us has followed the entire project focusing on a particular part of the entire design flow.

The three major skill areas were:

VHDL design: architecture definition of principal blocks starting from project directives, writing of VHDL code and functional simulation of every block, logic synthesis and netlist level functional simulation.

DFT Implementation and ATPG: DB-DAC Scan-chain insertion and ATPG (Automatic Test Pattern Generation), additional Test Logic Insertion in order to meet scan-chain Design Rule Constrains and testability directives.

Chip Implementation & Assembly: creation of the Layout view starting from Verilog netlist passing through all the intermediate steps such as: Floorplan, Placement, Special Routing, Clock Tree Generation, Routing, DRC/LVS checks and timing verification.

Obtained Results:

- *area decreasing of the DB_DAC block respects the original;*
- *implementation is of about 15%;*
- *over 95% Test Coverage;*
- *final routing successful with no timing violations.*

The new DB_DAC layout obtained with our new architecture is shown in Figure 3. We can notice the presence of 2 macro-cells instead of 3 as seen in the previous layout. In fact we used a single RAM and ROM instead of 2 RAMs and 1 ROM.

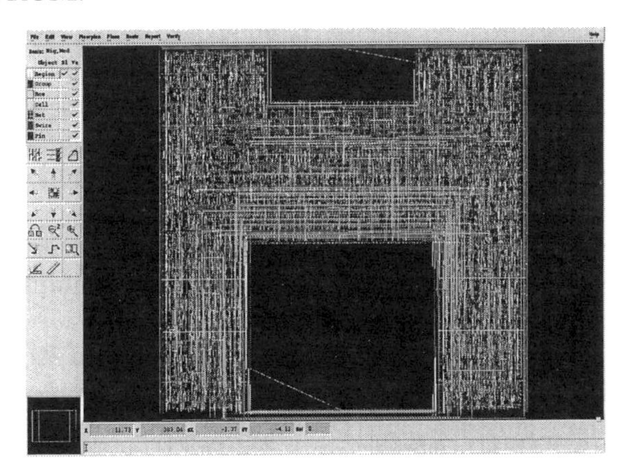

Figure 3: Layout of the DB-DAC module implementing our new architecture

VHDL Design, DFT & Layout Implementation Service Project

The Accent service provided in this project is conventionally named *VHDL Design and Layout Implementation*. This service is intended to implement a very close interaction between Front-End and Back- End activities.

The starting point is a set of Customer requirements. VHDL code is developed to implement the required hardware. This VHDL description is synthesized into a Verilog netlist that is verified running static and dynamic timing analysis (pre-layout in best and worst cases).

This activity checks for all possible problems generated after the process *RTL to synthesis* is completed.

The timing analysis allows the designer to define critical timing region that has to be kept tightly constrained, i.e., physically located close together blocks in the floorplan as well as clock distributions, insertion delay requirements and clock skew.

The same methodology is adopted for DFT (Design for Testability) issues like scan chain architecture, reordering and global test signals, BIST (Built in Self-Test) logic is confined to a region close to the respective RAM/ROMS. High test coverage for ATPG (Automatic Test Patterns Generation) is the main constrain in this phase.

The layout design is implemented starting from the same Verilog netlist used in timing analysis, this activity is composed by floorplan, clock tree generation and global routing, and requires a certain number of iterations to match the constraints. The primary constraint is to minimize the area.

Additional constraints are: the standard cell library to use, the number of clock and their skews, the power consumption and distribution, and the timing constrained nets.

Other informations that drive the layout design are: the knowledge of the design hierarchy, the flow of signals between blocks, the number of RAMs and ROMs, the input and output pins position.

The extraction of parasitic (*RSPF*, Reduced Standard Parasitic Format) from the layout allows to perform checks for timing violations in synthesis environment. The Figure 4 summarizes the explained flow.

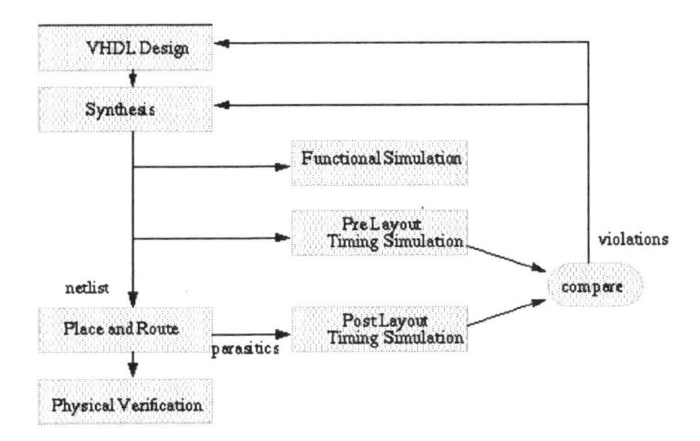

Figure 4: Typical flow of VHDL design & layout implementation service

VHDL Design of DB-DAC Module

The DB-DAC block functionality and the architecture used to implement it is described. Before we take a short overview on VHDL design methodology used in the design.

Some words about VHDL & synthesis

VHDL (*VHSIC Hardware Description Language*) is becoming increasingly popular as a way to design complex digital electronic circuits for both simulation and synthesis. Digital circuits designed using VHDL can be easily simulated, are more likely to be synthesizable into multiple target technologies, and can be archived for later modification and reuse.

In his introduction to *A VHDL Primer* (Prentice Hall, 1992), Jayaram Bhasker writes, *"VHDL is a large and complex language with many complex constructs that have complex semantic meanings..."*. This statement, with its possibly record-breaking three instances of the word "complex", reflects a common and for the most part correct perception about VHDL: it is a large and complicated language.

Once the designer has described the circuits in VHDL format he must check for his functionality writing 'Test-benches' and running simulations.

Test benches are extremely important to a complete design description. It is not simple describing the behavior of a circuit from the inside out. The only way to verify that a design description written in VHDL (or any other form of representation) operates as expected is to simulate it. Test benches provide the framework in which to perform such a simulation. The easiest way to understand the concept of a test bench is to think of it as a virtual tester circuit. This circuit, which is described in VHDL, applies stimulus to the design description and (optionally) verifies that the simulated circuit does what it is intended to do.

After the VHDL phase, the next step is to map the 'software' representation in the 'hardware' representation of the design through the synthesis process.

The design synthesis process consists in translating a RTL logic description into technology dependent building blocks. With the term RTL (Register Transfer Logic) we intend a subset of VHDL commands that are best suited for synthesis process; so RTL is the VHDL code to be synthesized.

This process can be very complex, depending on the design dimension, its structure, and the overall synthesis approach. Before starting writing any script (the file in which we specify all the synthesis commands), it is strongly recommended to have a global understanding of the design structure in terms of clocks, frequencies, critical paths, constraints as well as technology issues and conditions. Beside this, a complete constraints file must be filled together with the VHDL designer containing all constraints in terms of multicycle paths, false paths, input/output constraints, clock relationship, critical sections (timing, area, power) and so on [1].

DAC block description

The DB-DAC is a digital dual channel modulator device that receive as input 20 bit digital samples at a frequency rate of 44.1 KHz (CD Quality) for each one of the two stereo channel (Left/Right). The output is an oversampled digital signal that has to be converted by the analog block of the DAC.

A main benefit of oversampling process is a reduction in quantization noise, and an increase in *signal to noise ratio* (we obtain ~110 dB), over the audio bandwidth. This derives from a basic principle of converters stating that the total quantization noise power corresponds to the resolution of the converter and is independent of its sampling rate. This noise is, in theory, spread evenly across the entire bandwidth of the system. A higher sampling rate spreads a constant amount of quantization noise over a wider range of frequencies. Subsequent low-pass filtering eliminates the quantization noise power above the audio frequency band.

The DB-DAC is divided in top level sub-blocks; each of them is divided in even more different sub hierarchical level blocks. Top level blocks are the same both for the original Accent and our architecture implementation and are represented by the following figure:

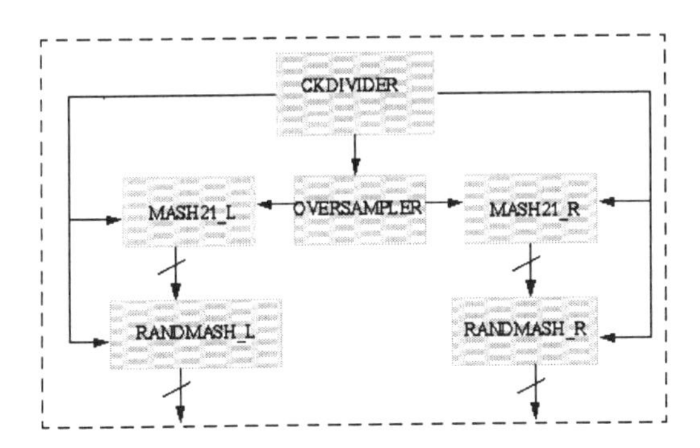

Figure 5: DB-DAC top level sub-blocks

A summary of the operation of the main blocks is indicated as follows:

CKDIVIDER: This block provides all the required clock signals for the DAC operation.

OVERSAMPLER: This block filters and oversamples the incoming data.

MASH21: This block modulates the 20 bit filtered data onto 3 bits. It is a MASH 2-1 structure (2nd order followed by a 1st order) Sigma-Delta modulator.

RANDMASH: This block used to reduce the noise in the signal by de-correlating the noise spectrum from that of the signal spectrum.

The principal block of DB_DAC is the OVERSAMPLER because is responsible of the oversampling through a particular algorithm called *convolution*. The convolution algorithm is realized with FIR filters that are mapped into hardware through RAMs, ROM, MAC (Multiply and Accumulate) and others modules together with a valid addressing and control logic.

The OVERSAMPLER block is the only one block that makes differences between our architecture implementation and the Accent's one.

Convolution algorithm implementation

Theoretical aspects that hide behind convolution algorithm will not be treated in this article, and we will only see which kind of operations OVERSAMPLE must carry on. All the below mentioned operation must be intended both for the Left and the Right channels of the incoming stream data.

We can see OVERSAMPLE as a cascade succession of 3 FIR Filters, each one with a different number of multiplicative constants.

Mathematically a FIR (Finite Impulse Response) filter with N+1 constants is defined as

$$y(n) = \sum_{k=0}^{N} c(k)x(n-k)$$

Where x (n) is the n-time input, y (n) is the correspondent output and c (k) are multiplicative constants. *The number and the value of those constants characterize the FIR filter.*

Another way to see at a FIR filter is to see it as a shift-register in which every register is multiplied by a constant and all the resulting addenda are added in order to obtain the output. For each raising edge of the clock signal, input enters the shift register and from each multiplicative line is obtained an addendum which will be added to the others in order to obtain the correspondent output.

If now we make follow to each real input a 'fictitious' input represented by a null sample, we can understand that with the same number of real input samples x (n) we can obtain a double number of output samples y (m).

The action we have just operated is called *oversampling.*

As we have previously said we can see the OVERSAMPLE module as a cascade of 3 FIR filter (with different constants) plus the addition of 3 oversamplers. In our design the first FIR is characterized by 84 multiplicative constants, the second by 24 and the third by 8. The number and the value of those constants derive from theoretical studies that we will not explain in this article.

For each convolution performed on FIR1, 2 convolutions must be performed on FIR2 (for the oversampler presence) and for each FIR2 output there are 2 correspondent FIR3 outputs. Every output of a convolution is memorized in RAM memories that are divided in three sections, one for each FIR filter.

The output of each filter convolution is written to the next section of RAM. The original architecture was characterized by 2 RAM memory blocks (one for each channel), each 128 words x 20 bit wide. The 128 words are further divided, by logical addressing, into three regions, one for each FIR filters.

The output of the first filter convolution (characterized by 84 constants) is written to the next section of RAM. The second filter (FIR 2) has 24 coefficients stored in ROM and unlike FIR 1, FIR 2 must operate on 2 data values. Therefore the FIR 2 algorithm must run twice. The results are then written back to RAM in new locations ready for the 3rd FIR. Finally, the third FIR filter (FIR 3) completes the oversampling and filtering of the input data. It has 8 coefficients stored in ROM and operates on each output generated by the FIR 2 (4 in total). Therefore the FIR 3 algorithm is run 4 times. This in turn produces 8 outputs. However, these are not written into RAM but to the SEQUENCER block. This block is basically used for re-synchronizing the signal data to the Sigma-Delta Modulator (MASH21 block). The filter algorithm writes data at a rate of 20 MHz but the Sigma-Delta MASH21 reads at a rate of 5MHz.

ROM & RAM mappings

All the constants are mapped in a single ROM and all the FIR samples are mapped in RAM(s). Each RAM is divided in 3 *section* and each of this section stores the samples of one of the 3 FIRs.

The most outstanding difference between our architecture implementation and the Accent original one is that we used a single double capacitance RAM for both the LEFT and the RIGHT channel instead of a single RAM for each of the 2 channels.

We made this choice because in general the addressing logic for a single RAM is less onerous (in terms of gate number) that for 2 RAMs.

The following figure explain better the logical memory mapping for FIR filter in each RAM as used in the original architecture:

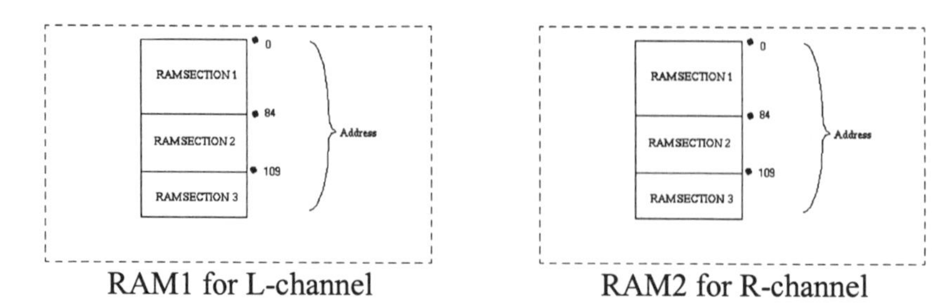

Figure 6: Old architecture

In our new architecture we have single 256 words x 20 bit wide RAM and a different mapping for each FIR filter and even for each channel.

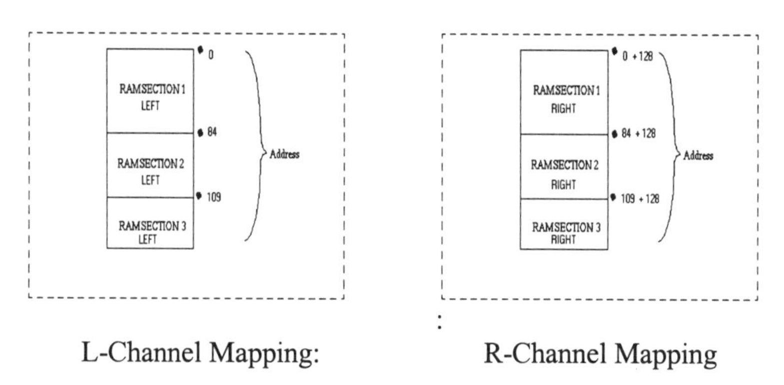

Figure 7: New architecture

So we leave an addressing bit for specifying which channel is under examination.

Our architecture of the OVERSAMPLER module

In the following page is displayed the top module schematic of our architecture for the OVERSAMPLE in the DB-DAC module (clock signal is omitted but it reach every block).

The principal differences between our and original architecture are:

-The use of a double capacitance RAM for storing both the Left/Right channels samples

-The presence of a single address generator for both RAM and ROM instead of RAM_L DECODER, RAM_R DECODER and ROM_DECODER of the previous implementation

-The presence of a SELECTOR for multiplexing the samples coming from the MAC or from the input

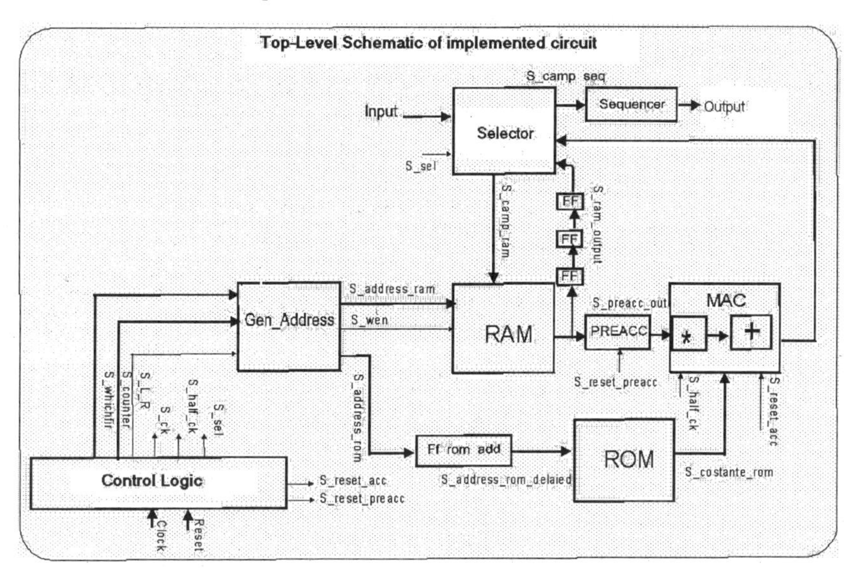

Figure 8: Top-level sub-block schematic view of our new architecture of module OVERSAMPLER

Lets now take a look of the main blocks of the architecture:

-RAM & ROM memory

RAM memories are taken from HCMOS6 ST-Microelectronics library and were generated ‘ad-hoc’ for meeting our requirements.

We used a single SPS3_256x20m4d16 that is a 256 X 20bits words, single access RAM. For storing the 20 bit FIR constants we used a ROM from the same technology library.

-Address generator

The main function of this module is to supply the correct addresses to RAM & ROM at the right time. The addressing logic is quite complex because it must take in account which FIR (and of which channel) is under examination and if is a reading or writing memory access.

-Pre-accumulator

This block sums a couple of his input and outputs the result of the sum.
It is evident that this block works at half the frequency of the master clock.

-MAC (Multiplier and Accumulator)

The MAC module is the heart of the OVERSAMPLE data elaboration process. The following figure shows the internal structure:

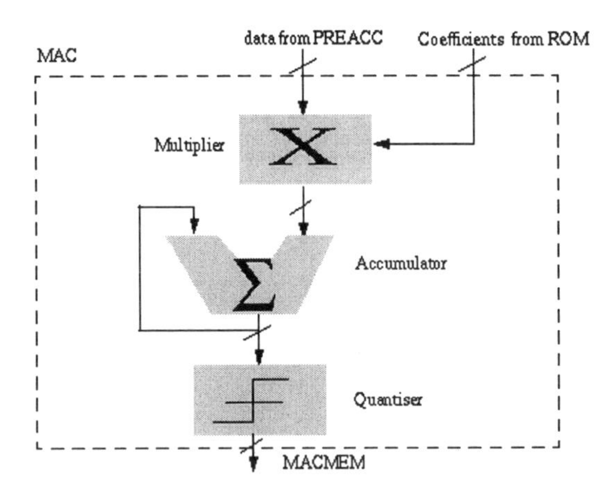

Figure 9: MAC structure

The internal block structure of the MAC consists of a 20x21 Multiplier followed by a 40 bit Accumulator and a 20 bit Quantizer.

At every *Half_Clock* rising edge the MAC module multiply the input received from the pre-accumulator with the correspondent ROM constant and then add the result obtained to the internal stored value and replace it.

The output for each *Half_Clock* rising edge is represented by the memorized sample during the previous cycle (*MACMEM*). A *reset* signal is utilized for resetting the internal register (*MACMEM)*.

-Selector
This module acts like a multiplexer because depending on which value is applied to *sel* input the sample input coming from convolution is passed to RAM or to SEQUENCER. Also, the OVERSAMPLER input samples are directly connected to this module.

-Sequencer
The need of the SEQUENCER block is basically motivated for re-synchronizing the signal data to the Sigma-Delta Modulator (MASH21 block). The filter algorithm writes data at a rate of 20 MHz but the Sigma-Delta MASH21 reads at a rate of 5MHz.

-Control Logic
This module is the 'brain' of the whole data processing because it synchronize in the right way all the modules of the architecture and it produce control signal for all the above mentioned modules. It behaves like a 'finite state machine' that at each clock rising edge passes from a state to another state. A *reset* signal is used for resetting this sequence of control signals.

DFT Implementation and ATPG of DB-DAC

In this paragraph we will illustrate the modifications performed on DB-DAC block in order to achieve a testable architecture. Firstly we will have a short introduction on testing issues and DFT methodology used in our design. Then a typical DTF flow will be described. In the last part, problems and results of our circuit will be reported.

Testing of integrated circuits

Testing is the production phase where the reliability of the integrated circuit is checked. All the circuits that get through the testing step are supposed to be manufactured correctly.

It is very important not to confuse this operation with verification task. In the verification phase it is checked that the design behaviour is the expected one. Obviously this happens before circuit manufacturing and is a completely separated step.

In this paragraph we are interested in the testing task and in what can be done, since design phase, in order to make a circuit easily testable. This design issue is called DFT, Design For Testability [2].

Fault models

What does mean that an integrated circuit is correctly manufactured ?
It means that not any physical defect is present. Thus during testing process every possible defect should be sought and relative integrated circuits should be rejected.

This operation is almost impracticable. For this reason the physical defects are grouped in classes, fault model, depending on the logic effect that the defect produce. Treating with fault models instead of physical defect simplify the problem but make hugely easier testing task.

Most common fault models are stuck-at-zero and stuck-at-one models. Each defect that results in a line stuck to logic value "zero" or "one" are included in this model.

Test pattern

How stuck-at models are used in testing? A well manufactured integrated circuit does not have any line stuck. Thus testing consist in checking if a stuck node is present.

How it can be done? Let suppose we want to ensure that a certain line is not stuck at *one-logic* value, we must carry this node to *zero-logic* using circuit inputs and then we must check, using outputs, that the node does not have the *one-login* value

In other words we have to find an input configuration that stimulates the node and at the same time let the node value propagate up to the outputs. This configuration is called Test Pattern.

At this point Testing issue may be reduced to the process of generating a Test Patterns set. It has to be minimal and should reveal the largest possible faults amount. In order to discuss this it is important to introduce Fault and Test coverage concepts.

Fault and test coverage

Fault Coverage is a measure of how good is a set of Test Patterns. It is given by the ratio between the faults that the set of Test Patterns can reveal and all the possible faults.

Test Coverage is more precise than Fault Coverage and takes into account the fault that intrinsically can not be revealed:

$$\text{Fault Coverage} = \frac{\#\text{ faults detected}}{\#\text{ total faults}}$$

$$\text{Test Coverage} = \frac{\#\text{ faults detected}}{\#\text{ total faults} - \#\text{ untestable faults}}$$

Automatic test patterns generation (ATPG)

Manual generation of a set of Test Patterns having an acceptable Test Coverage and with a small number of patterns is almost impossible. This operation is typically done by software programs called ATPG

These programs, reading a description of circuit, usually a gate level netlist, with a certain number of constraints and using very computationally expensive algorithms, generate Test Patterns. This task can have a great effort in term of time but it has to be performed once for a given integrated circuit.

Design for testability

Why should the design of a circuit be complicated by additional operations relating to Testing phase? A good ATPG program could seem enough.

The problem is that test patterns act only on external inputs and outputs. As a result, without Design For Testability, the only way to control the circuit are the inputs and the only way to observe it are through the outputs. Current circuits have few input-output ports in comparison with the large number of gates and this tendency is destined to grow.

In addition, largely digital circuits are sequential, that is, they have a status. ATPG need to control and observe this status. This makes the generation very difficult and the amount of required vectors, for a certain test coverage, increases sensibly.

This implies that some techniques should be adopted since the design phase, in order to obtain circuits that can be tested with acceptable effort.

Many methodologies have been developed. They include both rules to be followed in design phase and ad-hoc techniques. We will examine the most used methodology..

Scan-path methodology

With scan path methodology each memory element is substituted with a so-called *scan-cell* formed by the original memory element and some additional logic. This logic allows all the cells to be connected, during the test, serially together to form one or more chain, called *scan-chain*, of *scan-cell*s.

A *scan-chain* is a shift register that works only in test mode. It can be load and unload, writing and reading the state of part or whole the circuit. Each *scan cell* acts also as a virtual pin to control and observe directly inside the circuit.

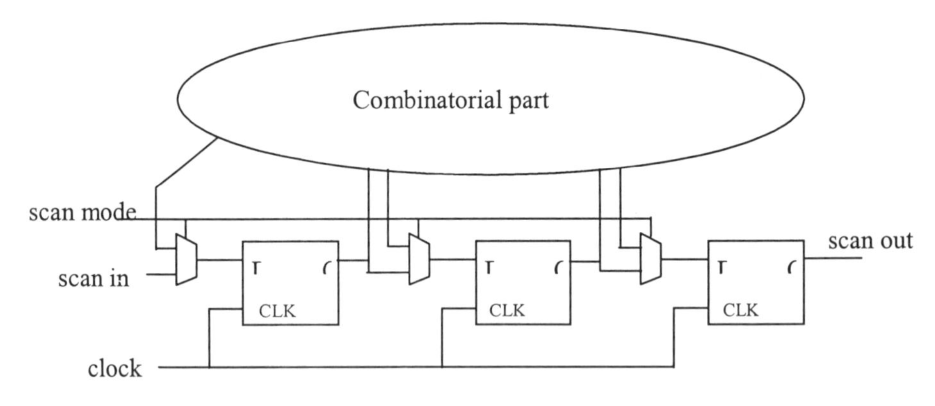

Figure 10: Scan chain insertion scheme

The Figure 10 shows a simplified picture of a circuit after the *scan-chain* insertion. Each flip-flop, that is each circuit that can store data, has been isolated from the combinatorial part. They are all connected in a shift-register configuration by multiplexers. A sequence of bits can be introduced and extracted serially through the input of first flip-flop and the output of the last flip-flop respectively.

Patterns generated by ATPG contains data for *scan chain*, too. Such patterns are longer but fewer and easier to calculate.

The price of the enormous advantage carried by this methodology, in term of testability, is an area overhead and a more complex routing.

Scan-chain insertion step is not a trivial task, some design-rules have to be followed to let the *scan-chains* work correctly. For instance, reset and clock of each *scan-chain* must be controllable, enable of latches must be controllable too, asynchronous loops have to be avoided, test clock have to be exclusive etc.

DFT Implementation flow for DB-DAC

Synopsys Design Compiler has been utilized for synthesis step. The scan chain insertion and Test-Patterns generation steps have been covered using both Mentor (DFTAdvisor, FastScan) and Synopsys (Test Compiler) tools.

Each flow has led to the same test logic insertion and to a very similar Test Coverage.

The implementation process is described in the following figure:

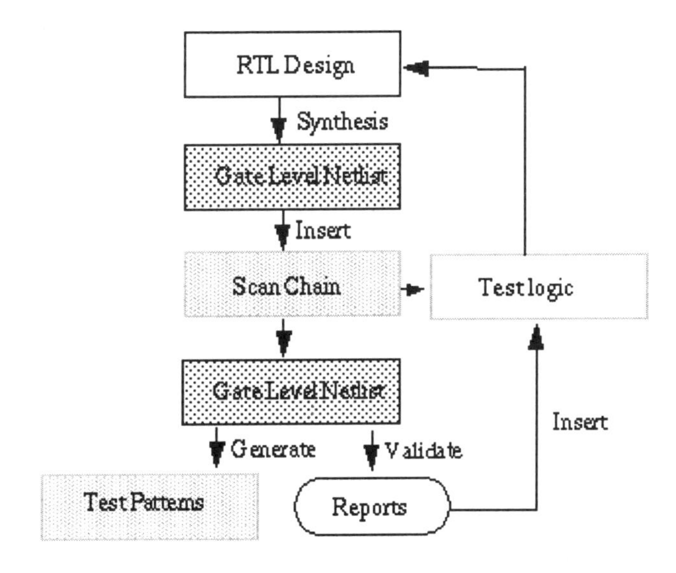

Figure 11: DB-DAC design flow

The process starts with an RTL description of the circuit. With the synthesis step a netlist is obtained. The *scan-chain* insertion step is performed on that netlist. Usually a feedback in the flow is needed. It results in a change of the RTL description or sometime directly on the netlist, consisting in a test point insertion. A test point is a modification aimed at increasing controllability or observability in a specific point of the circuit.

When all the *scan-chain* DRC (Design Rule Check) are meet *scan-chain* insertion step is completed. A new netlist, containing the *scan-chain*(s) is generated. ATPG is then performed on that netlist to estimate the Test Coverage achievable with an acceptable number of Test Patterns.

A second feedback is often needed at this level since the required Test Coverage should not be reached. This step consists in a test point insertion too. When the necessary Test Coverage is obtained Test Patterns are finally generated.

DFT Implementation and ATPG of DB-DAC

These were the specifications for test logic insertion:

- a test coverage greater than 95% has been required;
- the number of pins, added in scan chain insertion, had to be as small as possible (≤ 2);
- area overhead due to testing circuits lower than 10% of the original area;
- test logic introduced in order to achieve testability had not to conflict with clock tree generation task (CTGen step).

Scan chain insertion for DB-DAC

A full scan approach has been adopted with muxed scan methodology. The result has been a single chain containing 1383 *scan cell*s. A single clock for scan was used. All flip flops including those of the RAM and ROM BIST modules were placed in a scan chain.

In the case of the RAM/ROM memory BISTs, the clock needed to be bypassed through a MUX and inverted because the BIST works on the negative edge of CK.

The text below is the report of DFT Advisor scan chain synthesis step:

```
Total number of sequential instances    = 1513
Number of transparent latch instances = 130
Number of inserted scan chains          = 1
Number of new pins inserted             = 2
```

Each memory element, excepting 130 latches has been substituted with the corresponding scan cell as shown in figure:

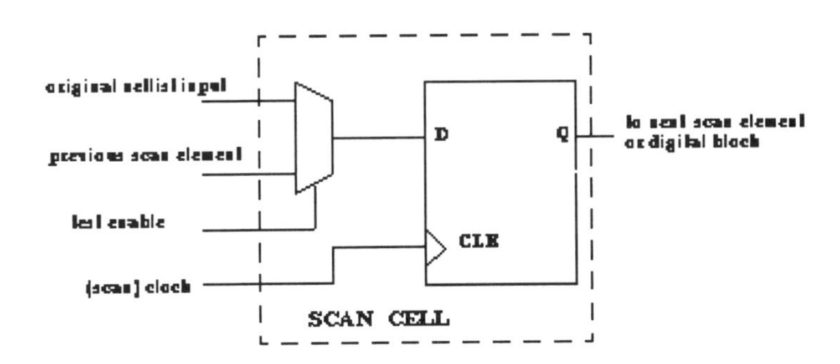

Figure 12: *Scan-cell* for DB-DAC

Test logic insertion for DB-DAC

The original DB-DAC design has been modified in several points in order to meet scan chain design rules and to obtain the required test coverage:

Test clock/reset management. Whenever a clock or reset is not directly controllable from a primary input, it is necessary to add some logic to control it. This solution is usually caused by internally gated clocks and/or resets or internally generated clocks and/or resets. The common solution is to insert bypass logic, two approaches, shown in the next figure, are possible:

- *Disable logic*: useful to bypass gated clock/reset structures disabling the gating effects on the clocks or forcing the reset to be always inactive in test mode. This solution reduces the ATPG coverage.

- *Control logic*: useful to bypass generated clock structures allowing a full control of sequential elements in test mode. It is usually the preferred solution.

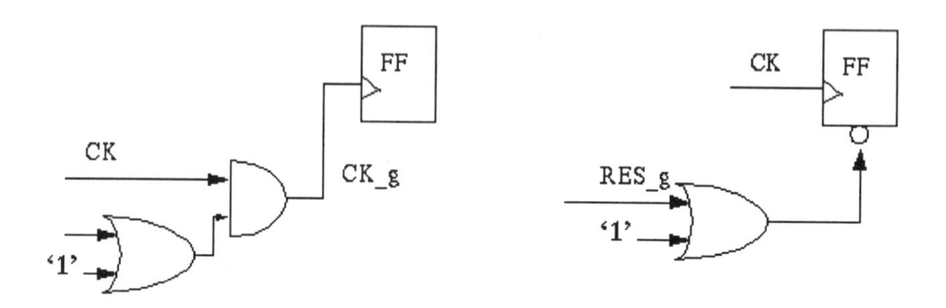

Figure 13: Control logic examples

As said before, in DB-DAC the global clock has been chosen as test clock. The other clock signals have been multiplexed with the test clock in order to use the same clock in test mode. CORELIB multiplexers were instantiated directly into the RTL code as it was considered to be the best means of connecting together the flip-flops in all the blocks with different clocks:

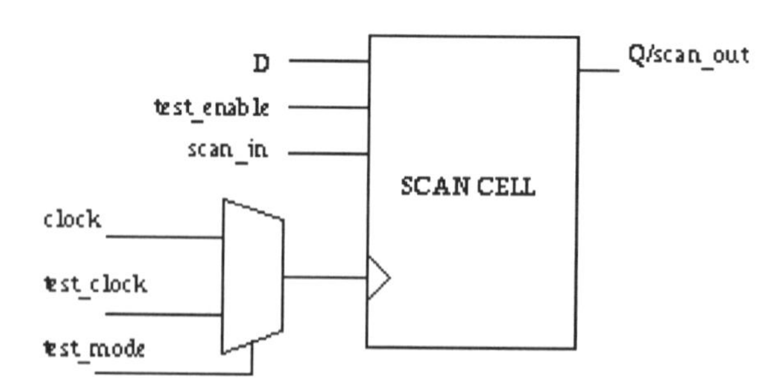

Figure 14: Multiplexer used to have a unique clock in test mode

A multiplexer instance is also strictly required to satisfy CTGEN compatibility. In fact, synthesis tool, in order to optimise occupation of area, may introduce not unapt logic (i.e., EX-OR gate) in clock path. This can make clock tree balancing seriously complex or even impossible.

In our project reset signal has been carried directly to the *scan cell* each time the reset terminal was uncontrollable. In the case described in the figure the reset signal is brought, through the test logic, directly to scan cell in test mode. Next figure shows an example of logic inserted in order to make a flip-flop with reset:

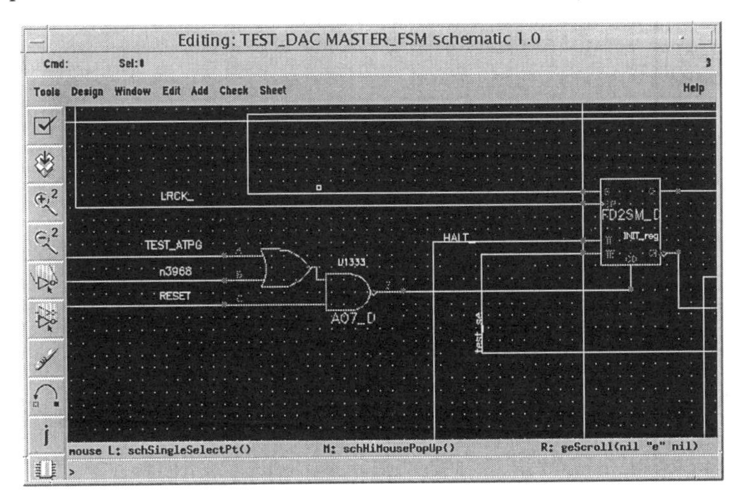

Figure 15: A typical controllability point for scan chain reset

Latch management. Latch usage must be controlled and confined to specific logic and timing needs because the methodology adopted requires flip flops to increase controllability and observability. Sometimes it could be necessary adding DFT control or disabling logic to keep latches transparent in test mode.

In DB-DAC were present 130 latches. They have been all recognised as transparent by the tools. This means that the tool was capable to control the clock pins of latches. Nevertheless, test coverage in the blocks containing those latches was unacceptable. Some logic has been added to set their clock terminals in test mode. In the figure, TEST_ATPG signal (the one before the OR gate) forces the clock input of the latch.

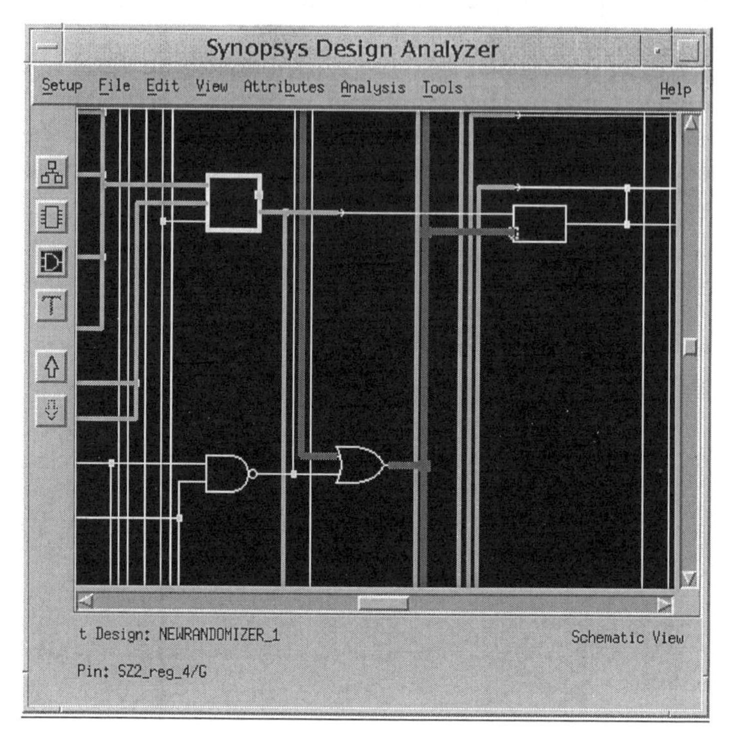

Figure 16: Control logic for transparent latch

Test pattern generation for DB-DAC

Test pattern generation has been executed on the resulting netlist. Statistics report shows the results obtained from the DFT and ATPG runs.

RAM and ROM blocks have been treated as black boxes. It will be noted that the inputs and outputs of the memory models were tied to X which means that they are not considered in the pattern generation and thus reduces the test coverage. However, the memory models are tested by their respective BISTs and thus are not excluded from testability.

All outputs have been considered observable. Detailed test coverage analysis has revealed that almost all the blocks in hierarchy has a coverage higher than 95%. Only the blocks connected with RAM/ROM blocks have a lower coverage.

```
Statistics report
------------------------------------------
                         faults
fault class              (coll.)  (total)
----------------------   -------  -------
FU (full)                  44083    68258
----------------------   -------  -------
UO (unobserved)               98      127
DS (det_simulation)        34531    56475
DI (det_implication)        7350     8338
PU (posdet_untestable)        82       85
UU (unused)                  322      442
TI (tied)                     63       78
BL (blocked)                   1        1
RE (redundant)               147      216
AU (atpg_untestable)        1489     2496
----------------------   -------  -------
test_coverage             96.26%   96.05%
fault_coverage            95.10%   95.02%
atpg_effectiveness        99.78%   99.81%
------------------------------------------
#test_patterns                        907
#basic_patterns                       899
#clock_po_patterns                      8
#simulated_patterns                  1312
CPU_time (secs)                      59.2
------------------------------------------
```

Chip Assembly and Layout Implementation

Now, the layout implementation flow utilised to implemented the physical chip [3] is shown.

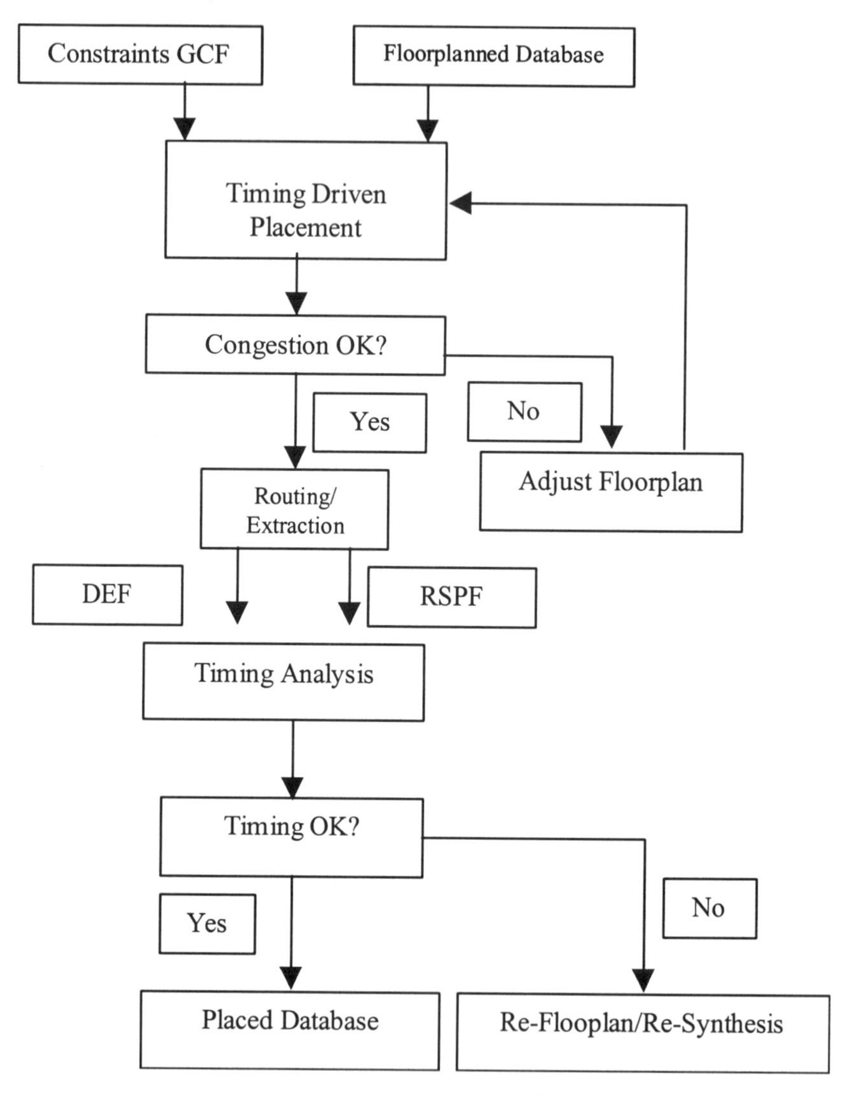

Figure 17: Chip assembly flow

General Constraint File (GCF) file provides boundary conditions and system level constraint information. The goal is to have consistent constraint usage and interpretation of semantics by all Cadence tools that support flow.

GCF specifies timing environment of the design, which includes clock waveform definition, arrival/departure time of input/output ports, slew rates for inputs, etc [5].

Verilog HDL is used to capture the architecture and the behaviour of a electronic systems. After logic synthesis a structural gate-level Verilog netlist can be created. Library Exchange Format (LEF) describes the library and technology information required for floorplanning and layout. Library data includes layer, via, placement site type, routing grids and macro cell definitions.

Design Exchange Format (DEF) is a representation of a raw netlist. The netlist describes instances of library components and sets of library model pins that need to be connected together.

The input to a floorplanning tool is a hierarchical netlist (Verilog or DEF) that describes the interconnection of the blocks the logic within the blocks; and the logic cell Floorplanning is a mapping between the logical description (the netlist) and the physical description (the floorplan).

The goals of floorplanning are to:

- arrange the blocks on a chip;
- decide the location of the I/O pads;
- decide the location and number of the power pads;
- decide the type of power distribution;
- decide the location and type of clock distribution.

In floorplanning is important to predict the interconnect delay before we complete any routing.

To predict delay it is necessary to know the parasitic associated with the interconnect capacitance (wiring capacitance or routing capacitance) as well as the interconnect resistance. At the floorplanning stage we know only the fan-out and the size of the block that the net belongs to.

After placement, next step is clock tree generation. To generate the clock has been utilised CT-Gen from Cadence. It is necessary to build a constraints file to pass to a CT-Gen tool and then analyse clock skew [4].

Clock tree generation flow

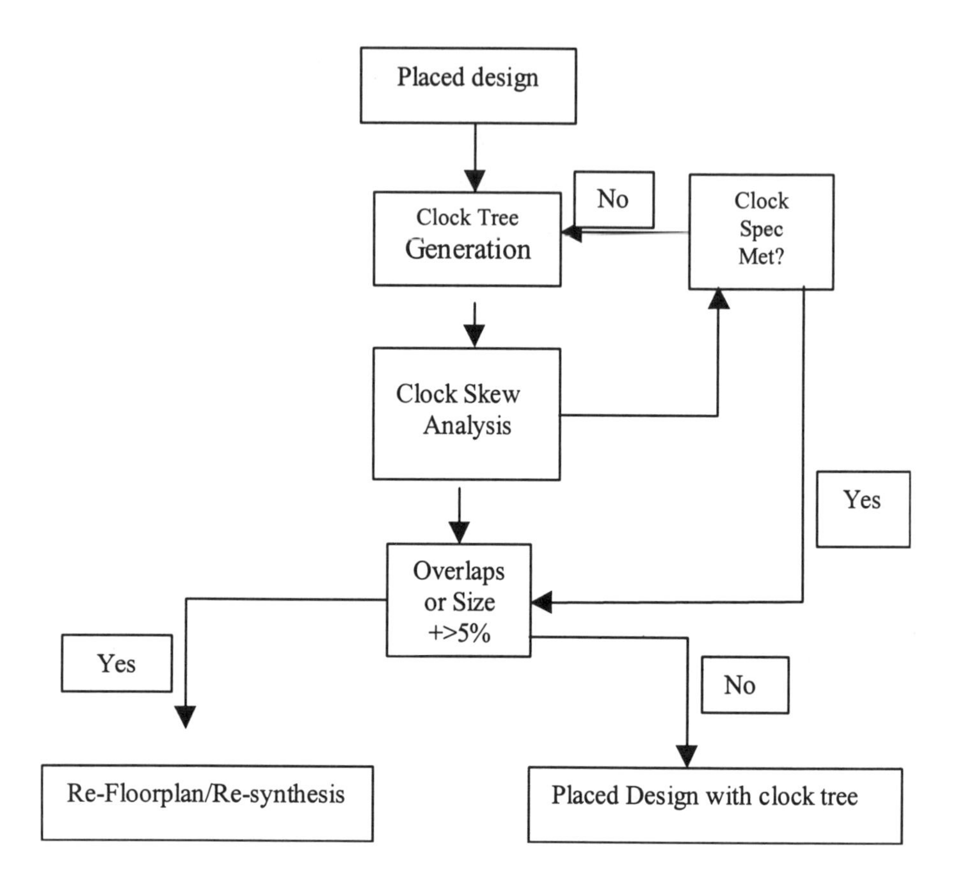

Figure18: Clock tree generation flow

Minimum delay is achieved when the taper of successive stages is about 3. We balance the clock arrival times at all of the leaf nodes to minimise clock skew [6].

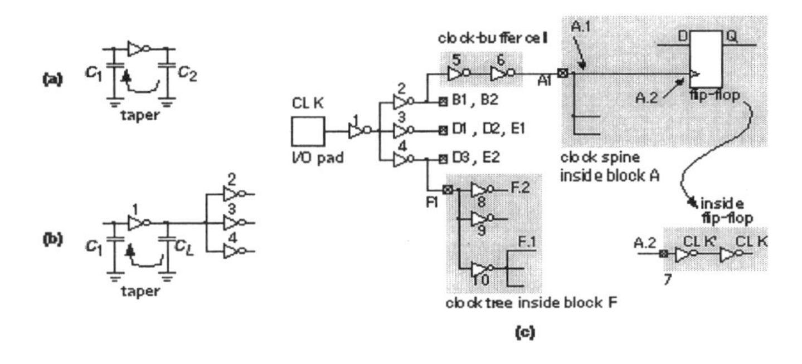

Figure 19: Minimum delay constrains

Clock tree is automatically built using CT-Gen of Cadence; following figure shows a simplified view of clock tree. The quality of the clock network-skew plays an important role in design performance. In fact, clock skew has a direct impact on a design's maximum operating frequency by reducing the time available for signals to propagate between storage elements.

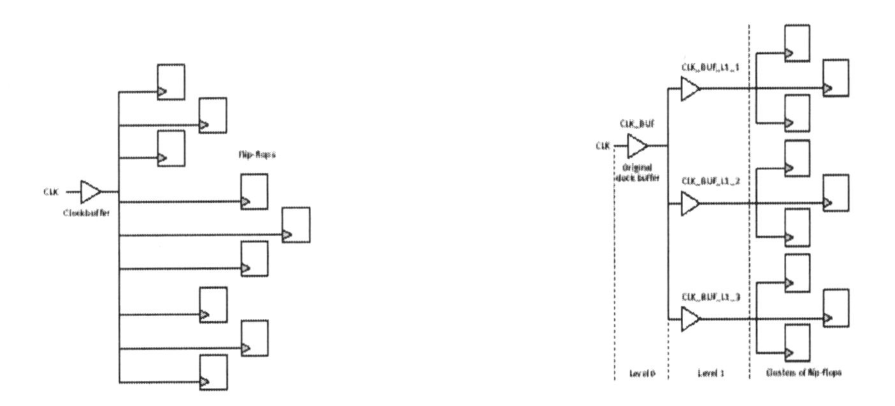

Figure 20: Clock tree before and after optimisation

Floorplanning and placement need a fast and easy way to estimate the interconnect delay in order to evaluate each trial placement; often this is a predefined look-up table. After placement, the logic cell positions are fixed and the global router can afford to use better estimates of the interconnect delay.

One method is that of the Elmore constant to estimate the interconnect delay for the circuit as shown in the following figure [7]:

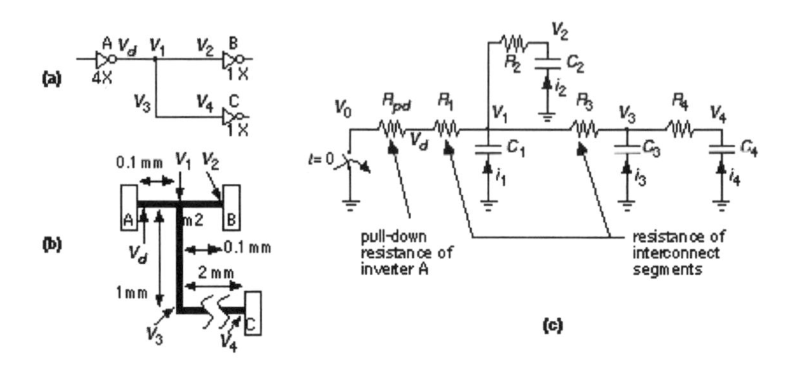

Figure 21: Elmore contestants estimation

The Elmore constant for node 4 (labelled V 4) in the network in figure (c) is:

$$\mathbf{t_{D4}} = \sum_{\mathbf{K}=1}^{4} \mathbf{R_{K4} C_K} = \mathbf{R}_{1,4}\mathbf{C}_1 + \mathbf{R}_{2,4}\mathbf{C}_2 + \mathbf{R}_{3.4}\mathbf{C}_3 + \mathbf{R}_{4,4}\mathbf{C}_4$$

where

$$\mathbf{R}_{1,4} = \mathbf{R_{p,d}} + \mathbf{R}_1$$

$$\mathbf{R}_{2,4} = \mathbf{R_{p,d}} + \mathbf{R}_1$$

$$\mathbf{R}_{3,4} = \mathbf{R_{p,d}} + \mathbf{R}_1 + \mathbf{R}_4$$

$$\mathbf{R}_{4,4} = \mathbf{R_{p,d}} + \mathbf{R}_1 + \mathbf{R}_3 + \mathbf{R}_4$$

Notice that $R_{24} = R_{pd} + R_1$ (and not $R_{pd} + R_1 + R_2$) because R_1 is the resistance to V_0 (ground) shared by node 2 and node 4.
It is necessary to perform two major checks before fabrication. The first check is a design-rule check (DRC) to ensure that nothing has gone wrong in the process of assembling the logic cells and routing. For this step two different tools have been utilised.

The Figure 22 shows the regions in which the design has been divided.

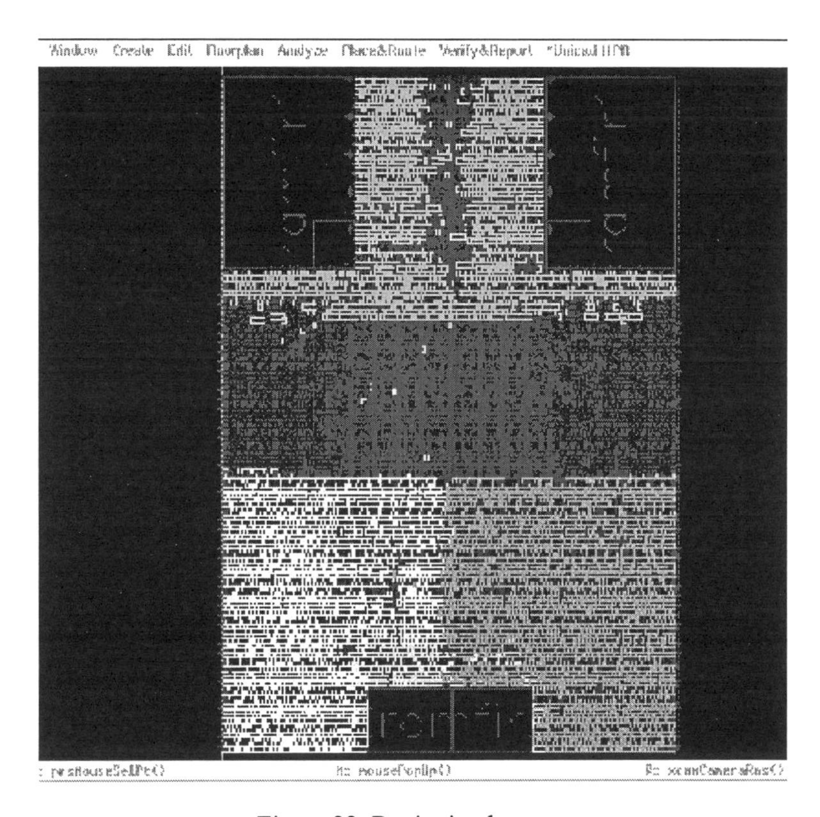

Figure 22: Regioning layout

- grey (upper) ram decoder and bist left and right;
- alu and fsm standard cells are distributed over the area;
- sequencer right and left just at the bottom of the two macro right and left;
- white: mash and randomiser left channel;
- green at the bottom right: mash and randomiser right channel;
- grey (lower) rom decoder and bist.

LVS results

The result of LVS run is summarised below:

****

********** LVS DEVICE MATCH SUMMARY *******************

****

NUMBER OF UN-MATCHED SCHEMATICS DEVICES = 0

NUMBER OF UN-MATCHED LAYOUT DEVICES = 0

NUMBER OF MATCHED SCHEMATICS DEVICES = 105874

NUMBER OF MATCHED LAYOUT DEVICES = 105874

****

********** DISCREPANCY POINTS SUMMARY *****************

****

32 DEVICE PARAMETERS (W/L/VALUE/AREA/PERI) MISMATCH

Conclusions

The project was successfully completed and some interesting results has been obtained compared to the original architecture. They are summarised below:

- area decreasing is of about 15 %;
- over 95% Test Coverage;
- final routing successful with no timing violations.

References

[1] Douglas L. Perry, *VHDL*, McGraw-Hill Inc. second edition.
[2] *Design-For-Test For Digital IC's and Embedded Core System*, Prentice Hall 1999.
[3] Sabin H. Gerez J., *Algorithms for VLSI Design Automation*, John Wiley & Son Ltd, 1998.
[4] Sadiq M. Sait, *VLSI Physical Design Automation: Theory and Practice,* Habib Yuossef, IEEE 1995.
[5] *Timing Driven Design Flow*, Cadence Manual.
[6] Eby G., Friedman, *Clock distribution Networks in VLSI Circuits and System*, IEEE Circuits and System Society.
[7] Neil H. E. Weste, K. Eshraghian, *Principles of CMOS VLSI design A System Perspective*, Addison-Wesley, 1993.

Improving the ST20C2P Microprocessor: An Introduction

Stella Arancio

Accent S.r.l.

Via F. Gorgone 6, 95030 Catania - Italy

E-mail: stella.arancio@accent.it

Introduction

The ST20 is a family of 32 bit microprocessors developed by STMicroelectronics for replying the raise request of microprocessor design for particular applications very cheaply and in a short time.

The aim of this work is to find the way to improve the ST20C2P microprocessor architecture. The meaning of improving is to simplify the port and the way in which it works in order to increase the working frequency.

In general, the way to improve the performance of a CPU is to reduce the critical path, where these are the paths of the CPU with the biggest delay. For back compatibility reasons the ST20C2P instruction set must be kept unchanged while trying to increase the performance of the single instructions. There are in fact some instructions that were linked with the transputer activity. Once the general structure of the processor is well understood, it is possible to propose some suggestions on the areas that can be improved.

The study of the single blocks begins with an accurate analysis from the micro architecture down to the Register Transfer Logic (RTL) of the sub blocks under examination, in order to understand the implementation and spot out the possible weak points. The fundamental step will be to analyse the report of the time analysis.

When the structure that introduces a critical delay in a path is found, it is possible to descend the architecture and change it. This means to study the Register Transfer Logic (RTL) of the block and to understand where is possible to introduce the modification.

The main purpose is to realise some embedded systems like DSP with high performance and high programmability. For this reason it is necessary to develop several CPU for particular applications. The current policy is to give to the client a specific product with the given constraints only with few changes done on standard design. In accordance with this policy, it is possible to improve and generalise the macro cells, in such way the "Reusability" and the minimisation of "Time to Market" are achieved.

The ST20 Microprocessor

The ST20C2 is a core processor [1] and any product using it will have a variety of different subsystems and peripherals integrated with the ST20C2. Some of the behaviour of the ST20C2 processor depends on these subsystems and peripherals. To fully understand the processor behaviour in a system, the behaviour of the subsystems must be understood.

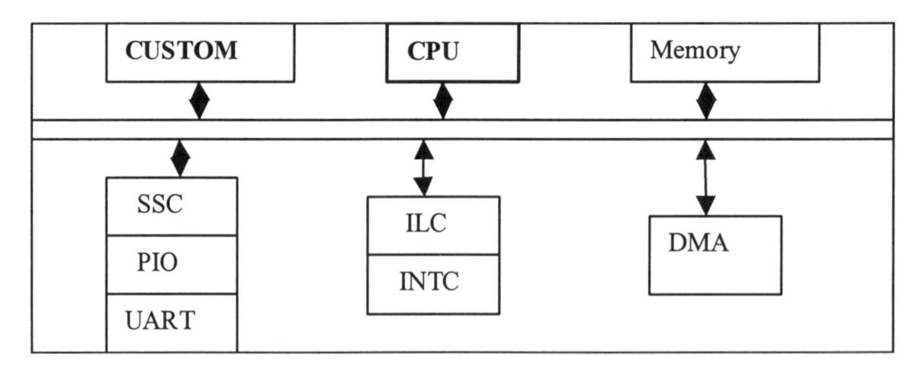

Figure 1: ST20 Configuration

The application fields of these microprocessors are various: they comprise Audio/Video applications, mobile phone and GPS. ST20 in general is composed by a CPU, Central Processing Unit, a DMA, Direct Memory Access, an EMI, External Memory Interface, a static RAM, an INTC, Interrupt Controller, a ILC Interrupt Level Controller and some peripherals like:

PIO, Programmable I/O, UART, Universal Asynchronous Receiver Transmitter, SSC, Synchronous Serial Controller.

According to the specific application, some of these parts can be changed or deleted for a custom design. In Figure 1 the typical configuration of the ST20 family is reported.

In Figure 2 a general scheme of microprocessor architecture is shown. It comprises the following main functional blocks:

- a CPU, Central Processing Unit;
- a Memory;
- some I/O;
- an interface to an interconnection network that connect it all.

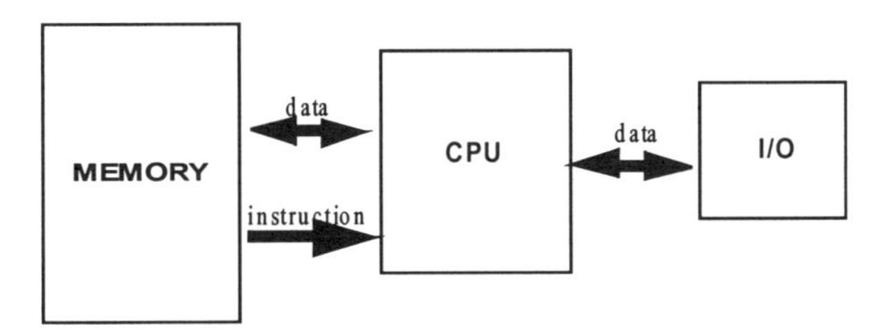

Figure 2: Basic architecture of a processor

ST20C2P is a particular RISC (Reduced Instruction Set Computer), in fact it has a very simple instruction set, making in this way the machine faster. Moreover, the instructions have variable length, in fact the complete acronym is "VLRISC", that is Variable Length RISC.

Further, in a CPU it is very important the type of internal storage, so in this section we will focus on the alternatives for this portion of the architecture. The major choices are a stack, an accumulator, or a set of registers. Operands may be named explicitly or implicitly. The operands in a stack architecture are implicitly on the top of the stack, in an accumulator architecture one operand is implicitly the accumulator and the General Purpose Register (GPR) architecture's has only explicit operands like register memory locations [2].

The major reasons for the emergence of General Purpose Register (GPR) machines are twofold:

- the first one is that the registers, like other form of storage internal to the CPU, are faster than memory;
- the second reason is that registers are easier for a compiler to use and can be used more effectively than other forms of internal storage.

The ST20C2 is a fast 32-bit microprocessor [1]. There are a number of advanced features able to enhance performance:

- a pipeline mechanism;
- a microcontroller;
- a mechanism of block-move;
- a trap-handling for error detection and debugging;
- an interrupt system to allow the processor to respond to external signals;
- an instruction fetch unit with 16 word register cache;
- a workspace cache which allows access to local variables to be overlapped with operations on their values giving effective zero-cycle access.

It has a real-time kernel; this means that the processor can run many processes concurrently. This can be used by an operating system, or with a language which has a model of concurrence, like occam or parallel C [1].

A single ST20C2 can efficiently implement a real-time operating system sharing processor time between concurrent processes. The on-chip real-time kernel provides fast context switch times, manages queues of processes at two priority levels, and handles the pre-emption of low priority activity by high priority one.

The ST20C2 provides mechanisms for caching a small number of frequently accessed memory locations and 'grouping' load operations from these locations with subsequent operations. This provides benefits both in terms of performance and in reduction of memory bandwidth requirements.

The most important and pervasive principle of computer design is to make the most common case fast. In making a design trade-off, the frequent case over the rare case is preferred. This principle also applies when determining how to spend resources, since the impact on making some occurrence faster is higher if the occurrence is frequent. Improving the frequent event, rather than the rare event, will obviously help performance, too [2].

For example, when adding two numbers in the CPU, we can expect overflow to be a rare circumstance and can therefore improve performance by optimising the more common case of no overflow. This may slow down the case when overflow occurs, but if that is rare, then overall performance will be improved by optimising for the normal case.

The register cache is a 16 word block of memory inside the ST20C2P which can be accessed immediately In general terms, this corresponds to the register file found in many microprocessor architectures, but with the critical difference that the registers act as a write-through cache shadowing the bottom 16 locations of the workspace structure in memory.

The register cache is used to accelerate the load-local instruction, which loads the n value from the workspace into the evaluation stack. A program is composed by more processes and each of them is loaded in the memory.

It is very important to remind the principle of locality: most programs do not access all code or data uniformly.

This principle, together with the guideline that smaller hardware is faster, leads to the well known hierarchy of memories of different speeds and sizes. Since fast memory is expensive, a memory hierarchy is organised into several levels. The goal is to provide a memory system which costs almost as low as the cheapest level of memory and runs almost as fast as the fastest level.

In applying this simple principle, designers have to decide which is the frequent case and how much the performance can be improved by making that case faster.

Each time changes the process and changes also the content of the Register Cache. If the process exceed the sixteen words then it is necessary to up date the *Wptr*.

The change of the context happens gradually. In a first moment to the context is cancelled, and in a second moment, when the process points new locations of memory, these location are loaded.

Top level

Two big functional blocks form the CPU: *IFetch* and Execution Unit, as in Figure 3.

Into these blocks, more simple sub blocks are contained. There are very dense interconnections between this blocks, in order to achieve a good performance of the CPU. The functions of these two blocks can be summarised as follows:

IFetch unit reads instructions from memory, four bytes at a time, into the Instruction Buffer; from here they are decoded, one or two bytes at time into an *opcode* plus operand, or a secondary *opcode* with no operand, which is passed to the execution unit. It is clear that the "I" on the *IFetch* word meaning Instruction, in fact Fetch Unit develop a important rule about the load and the decoded of instruction.

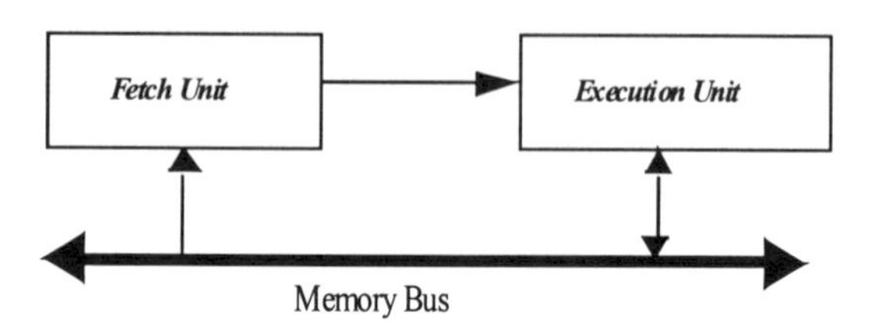

Figure 3: CPU blocks

The Fetch Unit is responsible for:

- reading out fetching instructions from memory (*IBuffer*);
- assembling compound instructions from the basic instruction stream (Predecoder);
- detecting errors and illegal *opcode*s (*ICheck*);
- passing the assembled instructions into the execution unit (Register Cache).

The *ICheck* module stands just in the middle between the *IFetch* Unit and the Execution Unit. It looks at the decoded instruction passed through the *opcode* bus for their "legality", that is, if they correspond to a legal or an illegal instruction. An instruction is considered illegal when its *opcode* is not one of the *opcodes* registered in the proper list, or when its *opcode* has been built from a sequence of more than 31 *pfix/nfix* instruction, causing the operand register to overflow. When an illegal *opcode* is signalled, a trap is raised.

The Execution Unit is responsible for execution of the instructions, managing the load for data, interrupts, traps and scheduling.

The Execution Unit can execute all non-prefix instructions, makes all the memory data accesses and contains all the ALU, scheduling, interrupt and trap mechanisms. It is divided into seven blocks:

- the ROM contains *microcode* for all the instructions, scheduling, interrupts and mechanisms;
- an Interrupt Arbiter which arbitrates between all the sources of interrupt, interrupts the *microcode*d engine and forces entry-points into scheduling and interrupt *microcode* routines;

- a Stack module containing *datapath* and control logic for the 3 word stack. The operand register and the logic, which supports pipelined memory reads, are also located in this module;
- an ALU module containing custom *datapath* for the arithmetic operations;
- the Hub module, which contains all the *datapath* and control logic for general purpose, machine state and functions;
- a Block Move module containing dedicated *datapath* and control logic for the *blockmove* operations;
- a Timers module containing dedicated *datapath* and control logic for the timer operations.

Instruction representation

This is a VLRISC CPU, for this reason there are:

- less use of memory for storing the programs;
- best interface with slow external memory.

The instruction encoding is designed so that the most commonly executed instructions occupy the least number of bytes. Now we describe the encoding mechanism and explain how it achieves this.

A sequence of single byte instruction components is used to encode an instruction. The ST20C2 interprets this sequence at the instruction fetch stage of execution.

A fundamental property of programs that is regularly exploited is locality of reference. Program tends to reuse data and instructions they have used recently.

A program spends 90% of its execution time in only 10%of the code. An implication of locality is that we can predict with reasonable accuracy what instructions and data a program will use in the near future based on its accesses in the recent past.

Generally we can say that the number of bit that composed an instruction is ten.

We have two types of instructions:

- “primary” instructions which have a direct;
- “secondary” instructions which have no operands.

The instruction encoding is designed so that the most commonly executed instructions occupy the least number of bytes. This chapter describes the encoding mechanism and explains how it achieves this.

A sequence of single byte instruction components is used to encode an instruction. The ST20C2 interprets this sequence at the instruction fetch stage of execution [1].

Instruction encoding

Each instruction component is one byte long, and is divided into two 4-bit parts. The four most significant bits of the byte are a function code, and the four least significant bits are used to build an instruction data value as shown in Figure 4.

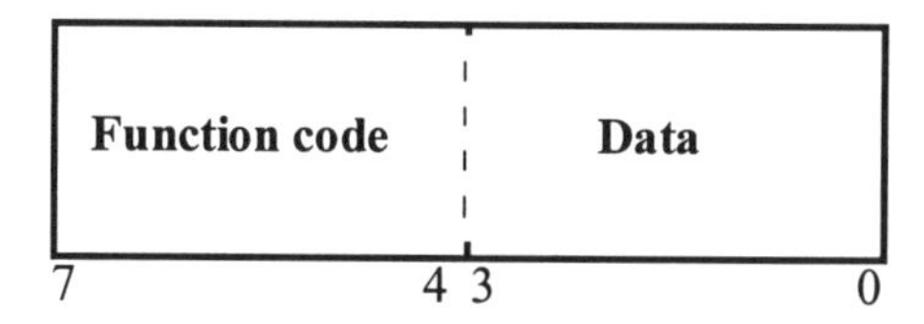

Figure 4: Instruction format

This representation provides for sixteen function code values (one for each function), each with a data field ranging from 0 to 15. Instructions that specify the instruction itself directly in the function code are called primary instructions.

There are 13 primary instructions. The other three possible function code values are used to build larger data values and other instructions. Two function code values, *p-fix* and *n-fix*, are used to extend the instruction data value by prefixing. One function code operate (*opr*) is used to specify an instruction indirectly using the instruction data value. *opr* is used to implement secondary instructions.

The instruction data value and prefixing

The data field of an instruction component is used to create an instruction data value. Primary instructions interpret the instruction data value as the operand of the instruction. Secondary instructions interpret it as the operation code for the instruction itself.

The instruction data value is a signed integer that is represented as a 32-bit word. For each new instruction sequence, the initial value of this integer is zero. Since there are only 4 bits in the data field of a single instruction component, it is only possible for most instruction components to initially assign an instruction data value in the range 0 to 15. Prefix components are used to extend the range of the instruction data value. One or more prefixing components may be needed to create the full instruction data value. All instruction components initially load the four data bits into the least significant four bits of the instruction data value.

The *p-fix* loads its four data bits into the instruction data value, and then shifts this value up four places. Consequently, a sequence of one or more prefixes can be used to extend the data value of the following instruction to any positive value. When the processor encounters an instruction component other than *p-fix*, it loads the 4-bit data field into the instruction data value. The instruction encoding is now complete and the instruction can be executed. The instruction data value is then cleared so that the processor is ready to fetch the next instruction component, by building a new instruction data value.

If more than 30 prefixes are detected then an illegal *opcode* trap will be signalled.

Primary Instruction

For efficiency, in the ST20C2 these are encoded directly as primary instructions using the function field of an instruction component.

Thirteen of the instruction components are used to encode the most important operations performed by any computer executing a high level language. These are used (in conjunction with zero or more prefixes) to implement the primary instructions. Primary instructions interpret the instruction data value as an operand for the instruction itself. These are instructions for loading and storing local workspace values only local variables, instructions for loading and storing bytes and 16-bit objects, and instructions for manipulating the values in the integer stack.

The most common operations performed by a program are loading and storing one among a small number of variables, and loading small literal values.

Secondary instructions

The ST20C2 encodes all other instructions, known as secondary instructions, indirectly using the instruction data value.

The function code *opr* causes the instruction data value to be interpreted as the operation code of the instruction to be executed. This selects an operation to be performed on the values held in the evaluation stack. This allows a further 16 operations to be encoded in a single byte instruction. The *pfix* instruction component can be used to extend the instruction data value, allowing any number of operations to be encoded.

Secondary instructions do not have an operand specified by the encoding, because the instruction data value has been used to specify the operation.

To ensure that programs are represented as compactly as possible, the operations are encoded in such a way that the most frequent secondary instructions are represented without using prefix instructions.

The encoding mechanism has important consequences:

- it produces very compact code;
- it simplifies language compilation, by providing a completely uniform way of allowing a primary instruction to take an operand of any size up to the processor word-length;
- it allows these operands to be represented in a form independent of the word- length of the processor;
- it enables any number of secondary instructions to be implemented.

Grouping

The grouping is a peculiar C2P mechanism, which allows us to describe this processor as a super scalar CPU. With the grouping the C2P can execute more then a single instruction per cycle, but without performing a parallel computation though.

The set of two instructions, which can be executed in parallel, is called "instruction group". They are executed at once and the global computation time is equal to the one of the slower instruction.

We cannot consider the C2P as a parallel processor since the two executed instructions are never processed in the same stage of the pipeline.

We can have a grouping by executing at once two primary instructions or one primary instruction and one short secondary instruction.

While the *ldc* instructs the CPU to load directly its operand into the A registers of the stack, the decoder 'decode' the add operation so that the microcontroller can generate the control signals to execute that operation.

The *ldc* instruction is immediately executed and can be considered to happen in a null cycle since it is contemporary to the add operation which last one cycle.

C2P Pipeline Description

Pipeline is an implementation technique whereby multiple instructions are overlapped in execution, it is the key implementation technique used to make fast CPUs. The throughput of an instruction pipeline is determined by how often an instruction exits the pipeline. Because the pipe stages are hooked together, all the stages must be ready to proceed at the same time, just as we would require in assembly line [1].

The time required between moving an instruction one step down the pipeline is a machine cycle. Because all stages proceed at the same time, the length of a machine cycle is determined by the time required for the slowest pipe stage. In a computer, this machine cycle is usually one clock cycle, sometimes it is two, rarely more, although the clock may have multiple phases.

The pipeline designer's goal is to balance the length of each pipeline stage, just as the designer of the assembly line tries to balance the time for each step in the process. If the stages are perfectly balanced, then the time per instruction on the pipelined machine, assuming ideal conditions, is equal to:

$$\frac{\text{time per instruction on not pipelined machine}}{\text{number of pipe stages}}$$

Under these conditions, the speed-up from pipelining equals the number of pipe stages, just as an assembly line with "n" stages can ideally produce cars n times faster. Usually, however, the stages will not be perfectly balanced; furthermore, pipelined machine will not have its minimum possible value, yet it can be close.

Pipeline yields a reduction in the average execution time per instruction. Depending of what you consider as the base line, the reduction can be viewed as decreasing the number of clock cycles per instruction (CPI), decreasing the clock cycle time, or as a combination of both. If the starting point is a machine that takes multiple clock cycles per instruction, then pipeline is usually viewed as reducing the CPI. This is the primary view we will take. If the starting point is a machine that takes one long clock cycle per instruction, then pipelining decreases the clock cycle time.

Pipeline is an implementation technique that exploits parallelism among the instructions in a sequential instruction stream. It has the substantial advantage that, unlike some speed-up techniques, it is not visible to the programmer [2].

The purpose now is to execute simple operations in each stage of pipeline, making is possible to increment the frequency up to 100MHz.

Once analysed the critical path of this CPU, it is clear that the most critical paths are those through the Add/Sub Unit and memory load and store.

The present pipeline structure can be subdivided into three stages: fetching, decoding and executing. If we want to apply the fore-mentioned criteria to this kind of structure we can think to improve the processor by modifying the first stage (in which the instructions and data are loaded from memory), or the third stage (in which the add/sub operations are executed as well as the storing actions).

Improvement

The new studied structure is presented in Figure 5. The five stages pipeline is needed in order to break the critical paths, thus arriving to a complete separation between the Fetch and the Execution stages. In this way we can increase the processor clock frequency since the operations to accomplish for each clock cycle are simpler and faster.

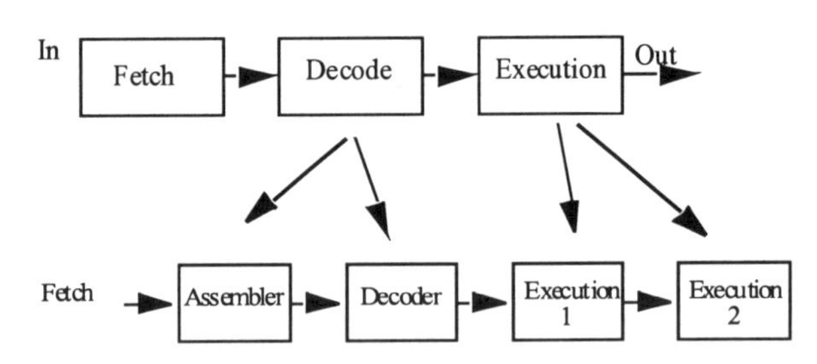

Figure 5: C2P pipeline

In this way the throughput (the number of operations executed by the processor in the time unit) can be increased, because when the pipeline is full the C2P could keep executing an instruction for clock cycle even if the clock period is much smaller [2].

By now we are focusing the resources to improve the first pipeline stage. We want to build a more efficient Fetch stage and then to begin to improve the other stages.

IBuffer

We already explained that one of the Fetch Unit blocks is the *IBuffer*. Its duty is to control the loading of the instructions from memory. We will simplify this block by taking off all the old and useless functionality that were used only by the transputer, thus we can have simpler interconnections and a simpler and faster control logic.

The *IBuffer* manages the eight bytes FIFO register that we already have analysed. The FIFO, First In First Out, is a register in which there is a queue to respect.

The structure of this FIFO must operate with two kinds of instructions (primary and secondary) and must take into account the loading from memory with a data bus width of single byte, half word and word.

Each possibility must be considered and every case has to be accomplished within the timing constraints.

The present implementation contains 24 multiplexers: seventeen multiplexers with two input ports, six with four input ports and one with three input ports. The proposed structure allows us to reduce the multiplexer number to two units, each of them with eight input ports. Of course, we refer to the number of multiplexer to handle the manipulations, so 8 must multiply the real number of physical multiplexers (one multiplexer for each bit).

The timing improvements that we aim to reach are related to the direct memory access accomplished by the use of eight registers, which will be renamed each cycle.

The instruction which must be executed passes by the multiplexer, which has eight selection signals. Those signals are needed to improve performance by coding them as 'one hot'. It is necessary also to check that the data present as an input to the multiplexer is valid. Effectively the data can 'be' or 'not be' present. The number of multiplexers must be doubled if we aim to load two instructions at once due to the secondary instructions which must be assembled in a second time by the decoder.

All the structure is represented in Figure 6. In this figure we can notice the new FIFO whit the bus for data <31:0> and for address <31:2> and the byte enable <3:0>. The Control block is relevant for managing operands.

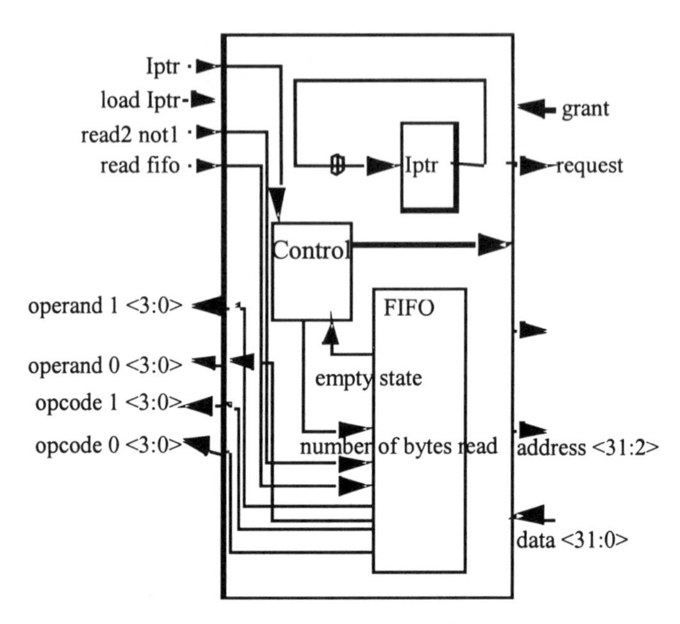

Figure 6: New *IFETCH*

Conclusions

The new structure should be faster because it is simpler. There are less blocks, but the more critical part is the control block. It is need a managing of trap and of grouping. In the Figure 6 the bit "read2 not1" is introduced: this bit is important to know if the current instruction is single or if is one of a grouping. If this bit is "1" the control must advance the FIFO of two bytes and load from memory two bytes more. In this block it is also important the interconnection with the bus. There is a particular protocol that allows us to read data from memory, these bits are "request" and "grant". These bits can update the FIFO.

It is interesting to note that new *IFETCH* must have the same bit in input and output like new FIFO, in this way it is possible to change only this block.

In any case, it is possible to increase the dimension of this new FIFO register without other fundamental changes and the performance of the microprocessor should be upgraded.

References

[1] ST20_C1, *Core Instruction Set Reference Manual.*
[2] Hennessy, J. L. and Patterson, D. A. *Computer Architecture: A Quantitative Approach*, 1990 Morgan Kaufmann Publishers, 1995.

Actual Status and Possible Development for CHIMERA Readout and Control System

Riccardo Papaleo

INFN - Laboratori Nazionali del Sud

Via S. Sofia 44, 95123 Catania - Italy

E-mail: papaleo@lns.infn.it

Introduction

One of the most interesting goals of the intermediate energy heavy ion research is to probe the properties of the nuclei under extreme conditions of density and temperature.

The hot and compressed system formed in the early stage of the collision can deexcite leading to multifragment final states. This multifragmentation is predicted to be the major decay mode for a nuclear system produced at high density and temperature [1].

An experimental investigation on the appearance of this process is of particular relevance to understand the basic properties of the equation of state (EOS) of the nuclear matter, but the occurrence of different reaction mechanism and the large number of nucleons involved in a reaction event.

CHIMERA is a project that attends of the physic study of the intermediate energy, so it attends the study of the collisions between heavy ions at the intermediate energy, this is a new field for the modern physics. The performances of the CHIMERA detector can then be summarised as following:

- angular coverage of almost (94%) 4pi;

- good angular resolution;
- identification in mass and/or charge of the detected particles;
- low detection threshold and high dynamical range in energy;
- direct velocity measurement for all the particles above the detection threshold.

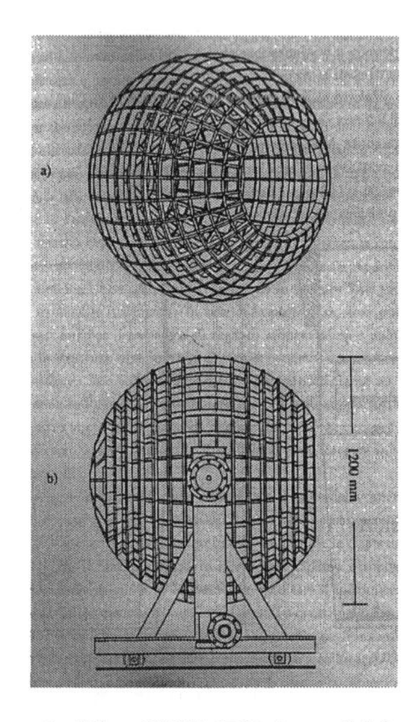

Figure 1: The CHIMERA multidetector

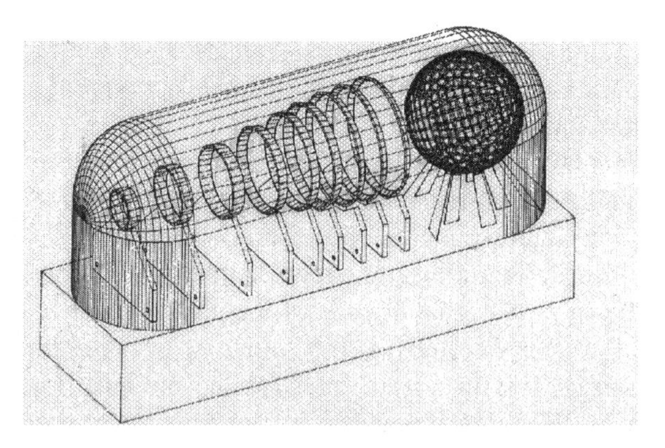

Figure 2: The CHIMERA multidetector

With this system we can measure the emission angle, the energy, the particle charge and/or the mass of every detected particle.

In the present work, the following topics will be analised:

- the status of the CHIMERA control and readout system;
- the characteristic of the principal electronic components of the CHIMERA control system (in particular the functional characteristic of the CAEN Model VN 1465);
- the DSP application on the project, in particular the WS2126 VME Sharc module;
- the possible development for the CHIMERA control and readout system;
- the development of a trigger system for the 4π detector;
- the functional characteristic of the integrate sensor.

The Device CHIMERA can be schematically described as a set of 1192 detection cells, arranged in cylindrical geometry around the beam axis, in 35 ring. The forward 18 rings are assembled in 9 wheels covering the polar angles between 1 and 30 are placed at a distance from the target variable from 350 to 100 cm with increasing angle.

The remaining 17 rings, covering the angular range 30-176 are assembled in such a way to shape a sphere 40 cm in radius. The CHIMERA detector uses different techniques to measure energy and velocity of the detected particles and to identify them in charge and/or mass. For particles stopping in the first stage, the silicon detector give an energy signal and a time information: velocity and energy are always measured, mass can be obtained using the time-of-flight technique. The detection threshold can be very low, depending only on electronics performances. For particle going trough the first stage, the combination of energy loss in the silicon and residual energy in the CsI gives the charge/mass identification, using the ΔE - E technique [1].

Energy and velocity are always measured. Light particles going trough the first stage without a detectable signal are identified in mass and charge in the CsI using the pulse discrimination technique (PSD).

The Detectors

A common problem in nuclear physics experiments is the charge Z identification of the particles emitted in the reaction with energy thresholds as low as possible. In general a two stage detector technique is adopted.

For instance, the latest generation of high resolution germanium arrays need efficient systems for particle detection tagging particular decay chains for superdeformation studies.These detectors must be as small and as light as possible, to avoid background generated by the compton scattering in the detector material. Low threshold silicon telescopes are a possible answer to these needs [2].

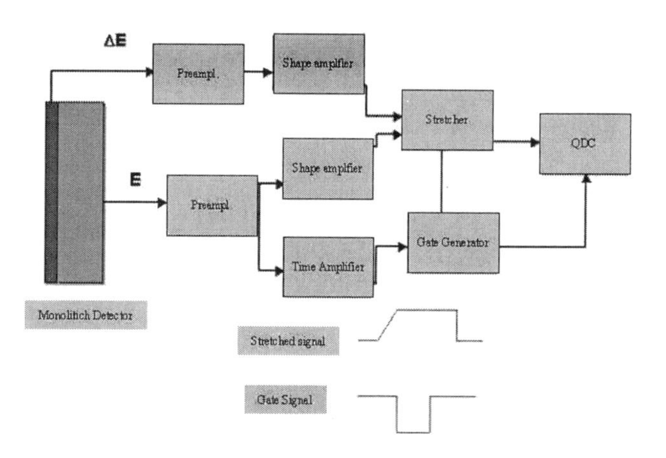

Figure 3: The acquisition system

To overcome the fragility problem, due to the small thickness of the first stage, and to reduce the costs, monolithic telescopes with the ΔE stage built over the E one had been proposed by many authors. An ultra-thin stage detector measures the energy loss of the particle ΔE, and a second stage detector measures its residual energy E. Combining the two signals, one can identify the particle charge, and in some cases also mass, using the energy loss dependence.

In order to obtain the smallest possible thickness for the ΔE quantity, the most powerful technique is probably to use high energy ion implantation.

Using this technique we developed an ultrathin monolithic telescope having a small area (4 x 4 mm2) and a strip detector with similar area for each strip.

Since larger area detection devices are often more suited in order to minimise the cost, decreasing the number of electronic channels, we developed a new larger with detector area; the E and ΔE stages are two reverse-biased diodes working at full depletion.

The buried anode collects all the holes produced by the ionising particles passing through the telescope; the upper cathode collects the electrons produced in the ΔE stage; the lower cathode collects the electrons produced in the E stage. This large area telescope (20 x 20 mm^2) was obtained modifying both in design and technology the older version area (4 x 4 mm^2).

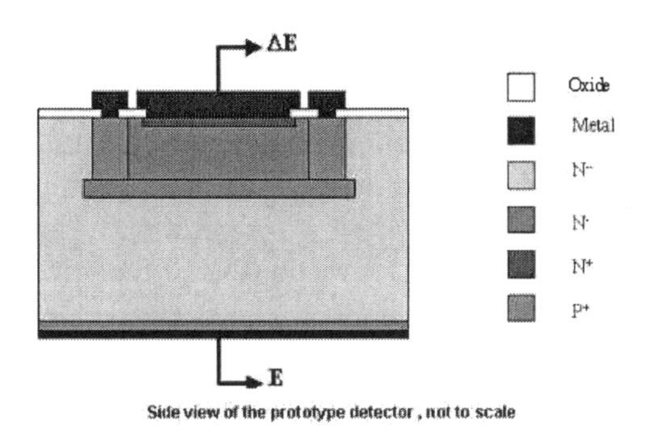

Figure 4: The silicon detector

The Cs(Tl) crystal

The high energy particles have the tendency to overcome the thickness of the silicon detectors easily, it has been necessary to use particular crystals, that with acceptable dimensions, furnished a strong stopping power to the high energy particles, moreover these crystals must be resistant to the damage provoked by the radiation and must give a good efficiency, for this reason the **CsI(Tl) crystals** have been chosen.

The crystals length has been calibrated as function of their stopping capacity of the high energy particles (protons). The Cs crystals are used to realise an isotopic discrimination of the detected particles, for this reason it was implemented a technique called "double gate".

Detector diagnostic

In a complex system such as the CHIMERA multidetector it is necessary to identify the fault condition with quickness, so it is necessary to have a system with good performances. In the CHIMERA detector, a control system in real time of the right working of all detector of CHIMERA it is realised with the use of DSP, **WS2126 VME SHARC** Cluster (ref. http://www.wiese.de/products.htm).

The DSP can be also used for the realisation of high level trigger, which is fundamental for the right analysis of the analog signals generated during the atomic reaction. The work is based on the spectral analysis of the signals, coming from all the detectors of CHIMERA. The signal is decoded and analysed by two **SHARC DSPs** (model 2160x).

The DSP has a very short response time (the execution time is about 25 ns), so it's adapted to analyse in real time the detected signal. Why we have chosen the DSP instead of an FPGA configuration? The principal reasons are: the cost, the flexibility, the time response and the prize

The Electronic Chain

The aim of the electronic chain is to acquire the analog signal coming from the detector chain and to analyse them. At the end of the analysis the data must be ready for the acquisition system. Moreover the electronic chain is used to supply the detectors [1].

The first electronic components, such as the pre-amplifier of the silicon detectors and the Cs detector, are placed inside the Ciclope's camera for minimising the effects of the parasitic capacitances and the lost power signal. The actual system has the required characteristics, but there's a high difficult to realise electronic components with the same characteristics, for this reason we are developing a VLSI system instead of the discrete one. The cabling of the CS photodiodes is realised with coaxial cables of the smallest possible length to minimise the noise. The silicon detector connection is realised with a soldering between the detector pins and the flat cables. The signals are conducted out from the camera through coaxial cables.

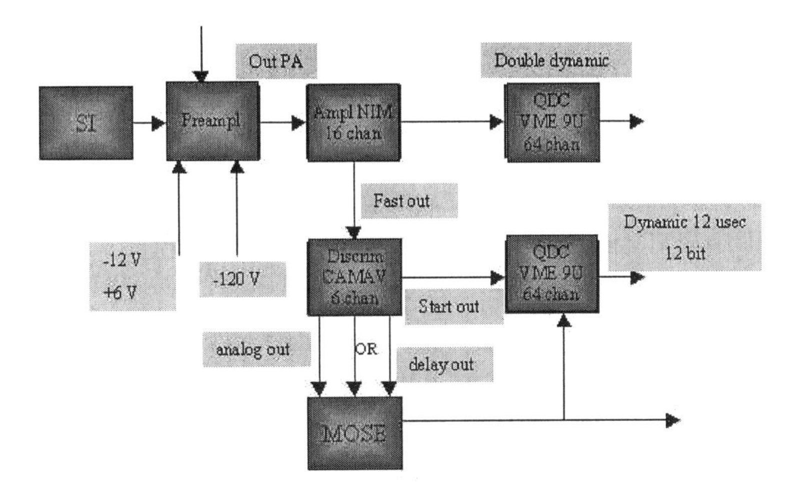

Figure 5: The electronic chain of the Si

The connectors, that are used for connecting the coaxial cables to the flanges, have been realised in particular way with the purpose to reduce the diaphony to the minimum.

The cabling has required a particular attention with the purpose to reduce the electronic noise to the least one, for such motive the cables disposition must be realised with a tree structure that prevents the generation of ring where the signals could bring to the generation of parasitic circuits and consequent electromagnetic induction [3].

The contacts are realised in a stable and sure way with the purpose to reduce every ulterior possible form of signal degradation to the least one. Finally, the problem of the mass has been analysed, the mass has to be unique for the whole apparatus, to avoid the generation of parasitic currents, such mass must be connected to the earth in independent manner and disconnected by every other mass that can bring external noise to the electronics.

The acquisition system

The system fulcrum is the Master Crate with the dedicated CPUs, used for the management of the data, that is coupled to the VME Crates where there are the A/D converter for the decoded of the energy and time (TOF) signals, the analysis systems and the data storage. The signals originate from the detected particles:

- time of flight (TOF);
- energy;
- rapid component of Cs detectors;
- slow component of Cs detectors.

Moreover there's a counter for the identification and the synchronism of all signals. It had been, considered a word of 32 bits of length for each parameter and for the counter address, and a maximum multiplicity of 50 particles has been token into account. These data are transferred therefore to the CPU that is occupied of the elaboration and of the following storage.

The system is controlled in real time with a visualisation system to histogram of the data coming from the detectors, over that from a new control system in real time of the correct detectors operation through cards DSP. It is used the technique of charge codify, consisting to integrate the detected energy in a temporal window (integration gate) well defined, the total ouptut value is the total charge quantity.

The Trigger System

The triggering system of a 4π detector must be able to perform an accurate event-selection task in order to enhance the visibility of the searched events.

The event rejection must be fast enough to avoid dead time. A further need for the CHIMERA trigger system comes from the choice of charge integration of silicon signals. Gates must be started as soon as possible in order to avoid any charge loss that could affect the resolution and the linearity of the energy calibration.

Using commercial QDC with common gate was in fact no possible the use of self-triggered gates as for instance used by INDRA. To open in time the gates we have to use a signal that must be available before the start of the signal from the shaping amplifier of silicon detectors (few hundred nsec after the start of the fast timing signal from a detector.

After the gate generation we have a quite large time to take the decision to perform the data conversion.

This time is the sum of the gate width (5 μsec) and up to 8 μsec that the used CAEN QDC VN1465 can wait for a clear operation, after the gate, before starting the true conversion [6].

Even if all this time can be used it is obvious that a large dead time can be generated in the case of a high rejection rate.

To lower this dead time, a faster decision would be preferred stopping the gate before its end and sending a fast clear to the QDCs. To satisfy such needs various levels of trigger are necessary. The MUSE is a UME module able to satisfy the above outline requirements. Its name MUSE stands for multiplicity selector. It will be used for the two level logic trigger selections and for the control of the whole readout system.

The required trigger levels are implemented starting from the Or signal of the detector and from the analysis of the geometrical multiplicity. External inputs are also available for both levels in order to assure enough flexibility for all experimental conditions especially for the coupling of Chimera with other detectors.

In the standard configuration the or signal of the whole detector will generate an inhibition signal window for the RF signal (zero level trigger) that will be, devolved after used, to generate all gates and stop signals to QDCs and TDCs with a good time reference.

At this level some simple coincidences can be performed with a coincidence window up to 100 nsec.

This zero level trigger will open our true trigger window mainly based on angular multiplicity measurements (first level trigger). The detector can be divided into 8 subsections, on each subsection a threshold of minimum multiplicity can be set and tested during the trigger window.

The 8 subsections are summed 4 by 4 into the 2 main sections where again the multiplicity is tested.

A further analog sum gives the total multiplicity of the detector. In this last stage two threshold can be set to allow us the test of a multiplicity window. A typical configuration can be used both to connect the CFDs (that generate the multiplicity signals) in order to give as first section forward rings of Chimera and as second section the sphere.

The selection of a high multiplicity in the sphere enhances the visibility of the most central events in heavy ion reactions. The more peripheral events can be easily excluded cutting events that give multiplicity only in the most forward rings [4].

Apart the angular multiplicity, up to 10 external inputs are available to generate the pattern of the event. This pattern is analysed by a fast memory after the trigger window and compared to predefined good configurations in order to accept the event. This allows, for instance, the insertion in the trigger of CsI detectors to take into account fast proton that does no fore silicon CFDs.

The maximum trigger window used will be about 2 μsec to avoid too much dead time, anyway to collect a good, even if no complete, multiplicity information, 500 nsec will be generally used a trigger window. After this time, if the configuration is rejected, all gates will be reset and a fast clear signal will be send to the converters.

If the event is accepted no action is performed, the system goes in busy up to the end of the conversion time because it collects and does the OR function of all busy signals from the different chains of QDCs—TDCs.

At the end of this conversion QDC and TDC are able to do a new conversion because they are equipped with a 16 events FIFO memory that can be read independently.

This FIFO is very useful to de-randomise the dead time.

Also MUSE system has this FIFO and saves the event pattern, the venet counter and an FDL pattern to be used for the readout as described in the following.

The CHIMERA converter readout is done by using the FDL (fast data link) system of CES. In each VME crate it is housed an FDL module that will read data from converters and will transfer them to the CPU in the master crate.

The FDLs need a trigger to start the readout this trigger is delivered by MUSE at the end of each conversion if the CPU is free or when the CPU become free if there are events to be read in the FIFO memory.

To each trigger can be associated a 4-bit pattern that will indicate the data destination.

This pattern can be used to store in different physical meaning (pulser, or other).

MUSE will allow the complete management of the FIFO memory, allowing a parallel readout and substantially decreasing the dead time of the ACQ that will be practically reduced only to the conversion time (100 μsec maximum).

The converter work with an integration gate for the QDC and a time signal for the TDC, these signal are sent together to the MUSE with a time command (strobe) chosen as reference signal.

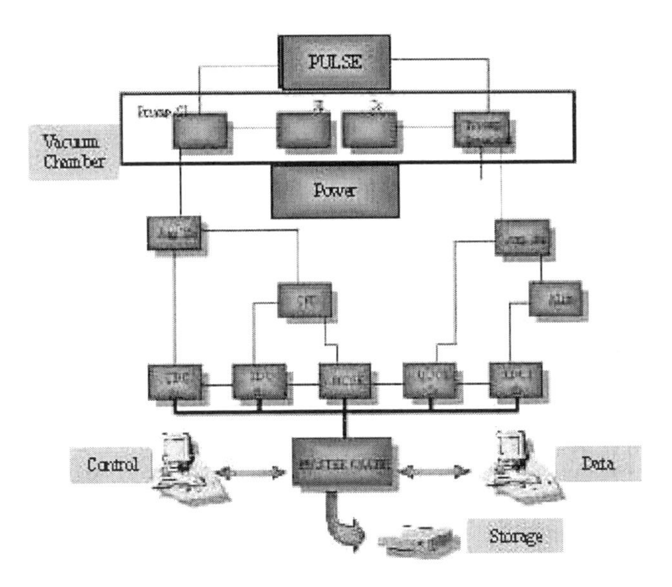

Figure 6: The electronic chain

At this point the start conversion is equal for all the converter, independently from the event.

If the event is not correct, after a comparison with the data storage in the memory, the module stops the gate and the start and stop signals, and it sends a reset signal (fast-clear) to the converter, that in 500nsec prepares the system for a new acquisition.

If the event is correct, thus the system generates the gate signals and the signals for the TOF measure.

The time is in a range of 5 μsec - 100 μsec, the system generates a *QDC-busy* signal, so that the CPU is busied for the elaboration of the event. After the processing and the codification phases, the data are sent to the CPU in FDL standard, the CPU generates a *CPU-busy* signal therefore the CPU is not ready to accept other valid data.

The analysis time of the CPU is about 100 μsec, this is a dead time for the other valid events, because the CPU is not ready to process other information, but the system is not in a dead time, in fact the QACs have memory that permits the storage of 16 events.

Clearly, it's possible that the CPU is busied and the MEB is full, so the QACs generate *QDC-busy* signal that does not qualify the sent off gate, start and stop signals. In this way, no events can be acquired.

This is the worst case, in fact in this time the system is dead. More short is this time, more efficiency is the system.

The maximum rate of the system, with this logic, is a function of a lot of parameters, but the maximum range is 1.4Gb, that corresponds to a data flux of about 1200 events / sec. The estimate of the dead time is correlated to the management time of each detected event.

This time is equal to the conversion time because the transfer e memory process come together in parallel.

In this way, the MUSE does not stopped the codifiers, those continue to work with the output memory

The system can reach a data flux of about 1Mbyte/sec, that means a rate of 800 events / sec, so we have a dead time of 20%.

The choice of the right events through the use of the system Muse will depend on the typology of the experiment in progress. The Muse is therefore a system able to effect a rapid selection of the events founded on the reconstruction of their multiplicity.

The first stage of the Muse is constituted by an element that we could define as a multiplicity assembler, the system doesn't do anything else than to take multiplicities (or events) analysed by the detectors and to give a string to a generator of pattern, these generate a logical configuration (pattern) related to the multiplicity of the event in examination.

The generated pattern is compared with the data inside the ECL memory (24 bit memory), this memory contains the desired pattern configuration for the experiment in progress.

If the data satisfies the desired configuration, the system sends it in out in FDL format.

The data is stored in a external memory for a following analysis. The "event counter" signal is combined with the "End CPU Busy" signal and processed by the FDL controller that generates the "FDL trigger" signal.

If the pattern does not satisfy desired features the system clear the data and generates a "fast clear" signal that resets all the converters for a new acquisition.

The working of the MUSE is a function of a combination between the time signal coming from a detector and the logic OR of all discriminator.

The signal is forbidden by the QDC-TDC busy signal.

This signal indicates that the detected event satisfies the desired features, or indicates that the CPU is busy at the time [5].

The timing circuit

The circuit for the management of the MUSE timing is in progress.

The new idea is the possibility to realise the same circuit in VLSI technology, so to realise a chip used for the management of the signals and the timing of the process.

In the management of the timing signals can have two operative condition:

The signal detected by the silicon and Cs detectors have furnished a satisfactory pattern.

The pattern, furnished by the detected signal, is outside from the user features [5].

Right signal

The signals are acquired from the silicon and Cs detectors, then they are amplified and shaped by dedicated amplifier.

After this step, they are sent to an electronic block, the "*Multiplicity Builder*", this block reconstructs the multiplicity of the detected signals.

These signals are used to reconstruct the pattern of the detected event. The detected signal are not necessarily synchronous, so they can arrive to the pattern into latch or registers the signals until the detected signal sequence is ultimate, unit in different time window.

If we want to reconstruct the detected event, it is necessary to memorise an external signal, called "Coincidence window", qualifies the latch to give the data in output, these data, furnished in numeric string format, are compared with the data inside the memory (these are experimental data).

If the elaborated data are equal to the data in memory, the "coincidence window" signal is sent to the signal generator that qualifies the four gate signals fundamental for the detectors [5].

The generated signals are:

- QDC signals for the silicon detectors;
- TDC signals for the silicon detectors;

- Fast signals for the Cs detectors;
- Slow signals for the Cs detectors.

The signal features of the Gate, for the different detectors, are shown in the Figure 7 and Figure 8, and they are a function of the detector features and of the electronics components of the chain.

The gate generator generates another signal that is used to increase a counter, this counter considers all the right events detected during the experiments.

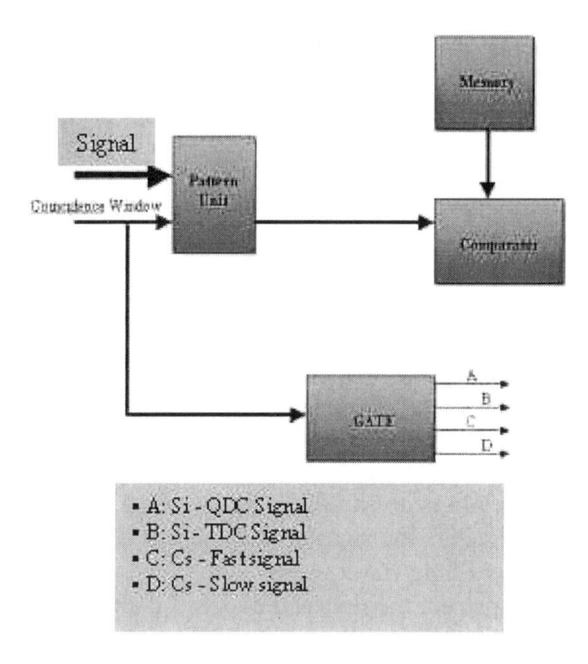

Figure 7: Right event

Wrong event.

If the event is wrong is necessary to reset all the detector and the gate signals, this is important to avoid useless data conversion, that could give wring information to the analysis system.

The only signal, that is emitted before the comparison between the data pattern and the data inside the memory is the gate signal of the silicon QDC (this signal has delay zero as shown in the following figure), so it's necessary that the comparator generates a "*clear generator*" signal that permits to the Clear Generator to reset all the QDC end to stop the generation of other gate signals.

All the gates will be resetted, and the counter will not be incremented.

The system will be ready to start a new acquisition phase.

The clear signal must start inside 50 ns form the pattern comparison.

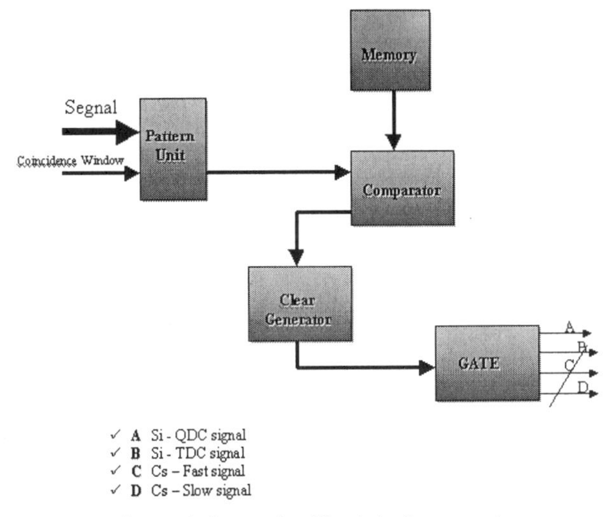

Figure 8: Wrong event

Conclusions

In this work the actual status and the possible developments for the CHIMERA project have been shown. The VLSI integration of the analog electronic and the introduction of non-linear control system such as neural networks, fuzzy logic, or other system will improve the performance of the multidetector device.

References

[1] G.Politi "CHIMERA: un rivelatore 4p per particelle cariche per lo studio delle collisioni tra ioni pesanti alle energie intermedie", PHD Thesis.
[2] S.Tudisco, F.Amorini, G.Fallica, G.Cardella "A new large area monolitich silicon telescope", Nuclear Instruments & Methods in physics research (1999).
[3] A.Musumarra, F.Amorini, G.Cardella, P.G.Fallica "implanted silicon detector telescope: New developments", Nuclear Instruments & Methods in physics research (1998).
[4] A.Musumarra, F.Amorini, G.Cardella, P.G.Fallica "A monolitich silicon detector telescope" Nuclear Instruments & Methods in physics research (1996).
[5] S. Aiello, A.Anzalone, G.Cardella "Muse: A trigger system for the CHIMERA multidetector.
[6] Technical Information Manual "MOD. VN 1465 S", CAEN.

Passive Component Modelling with HFSS

Antonino Scalambrino

Accent S.r.l.

Via F. Gorgone 6, 95030 Catania - Italy

E-mail: antonio.scalambrino@accent.it

Introduction

Radio frequency (RF) circuits fabricated in monolithic integrated circuit technologies make extensive use of on-chip transmission lines for the realisation of inductance, which are key components in many high-performance narrow-band circuit designs.

The transmission line widely used for high-frequencies (from a few hundred megahertz to over 10 GHz) is the microstrip line, because of its simplicity in geometry and simplicity of construction. It consists of a metal strip above a conducting (or ground) plane with the substrate and intermetal oxide layers sandwiched between two conductors.

In order to exploit the capabilities offered by a monolithic inductance, constraints on the component performances due to the silicon technology must be accurately modelled.

An accurate characterisation of this structure at microwave frequencies requires an analysis of the fringing fields, parasitic, round plane effect, and an analysis of the conductive substrate effect on the component performance, being the last one an issue in silicon IC design. These effects cannot be fully analysed to yield closed-form expressions, which adequately predict the inductor behaviour, and hence numerical analysis is required in order to determine the parameters of the inductor's electrical model.

On the basis of the above consideration, simulators allowing the analysis of electromagnetic fields are used. The HFSS simulator, developed by *Ansoft*, has been taken into account. In particular, in this contribution the performance comparison between this tool and to two-dimensional simulators actually used is addressed.

To estimate the performance results of a tool it is necessary to compare them with accuracy measurements carried out in laboratory on sample elements.

A metal3 microstrip with 1mm length, 10 μm width and 1.5 μm thickness in HSB2 (High-Speed Bipolar) technology has been implemented, because of the availability of measurements on this device carried out in the R&D electronic laboratory.

During project development several difficulties, not all overcome, have been found to define the structure ports.

Although the passive component characterisation points out that the model used is valid, a numeric disagreement between measured data in the laboratory and simulated data has been observed. This is due to both the dependence of S-parameters on the uncertainty in the ports dimension and to the different definition of the boundary conditions in a multi-layer structure.

HFSS—Electromagnetic Simulator *Ansoft*

The simulator HFSS, High Frequency Structure Simulator, is a software developed by the AnSoft, allowing S-parameters calculation of passive high-frequency structures, such as microstrips, waveguides and transmission lines. In order to generate the electromagnetic field solution this tool uses the Finite Element Method. The simulator also includes post-processing command for analysing the electromagnetic behaviour of a structure in more detail.

Finite Element Method (FEM)

The Finite Element Method divides the full problem space into thousands of smaller regions and represents the field in each sub-region (element) with a local function. The geometric model is automatically divided into a large number of tetrahedral, where a single tetrahedron is basically a four-sided pyramid. This collection of tetrahedra define the mesh of the structure.

The value of a vector field quantity (such as the H-field or E-field) at point inside each tetrahedron in interpolated from the vertices of the tetrahedron. At each vertex, HFSS stores the components of the field that are tangential to the three edges of the tetrahedron. Moreover, the system can store the component of the vector field at midpoint of selected edges that is tangential to a face and normal to the edge (as shown in Figure 1). The field inside each tetrahedron is interpolated from these nodal values.

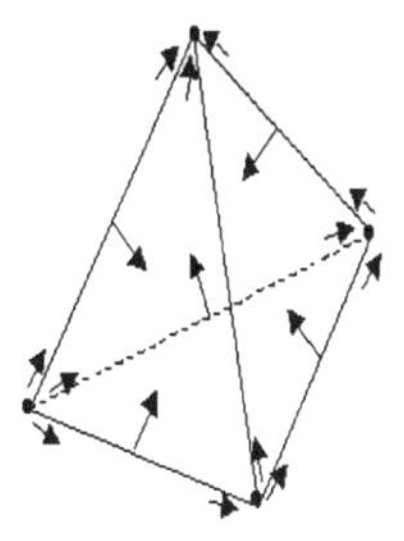

Figure 1: Representation of a field quantity

By representing field quantities in this way, the system can transform Maxwell's equations into matrix equations that are solved using traditional numerical methods (successive approximation).

There is a trade-off between the size of the mesh (the desired level of accuracy) and the amount of available computing resources. The accuracy of the solution depends on how small each of the individual elements (tetrahedra) is. Solutions based on meshes using thousands of elements are more accurate than solutions based on coarse meshes using relatively few elements. To generate a precise description of a field quantity, each element must occupy a region that is small enough for the field to be adequately interpolated from the nodal values. To produce the optimal mesh, HFSS uses an iterative process in which the mesh is automatically refined in critical regions.

First, it generates a solution based on a coarse initial mesh, then it refines the mesh in areas of high error density and generates a more accurate new solution. Mesh definition for a three-dimensional antenna is shown in Figure 2. It should be observed that the dimension of tetrahedra are smaller in the critical regions than in the other regions.

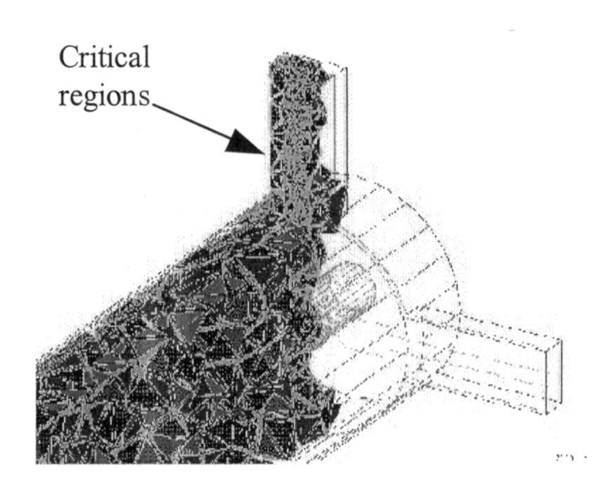

Figure 2: Three-dimensional structure mesh

Design Flow with HFSS

A complete design with HFSS need a orderly steps sequence as shown in Figure 3.

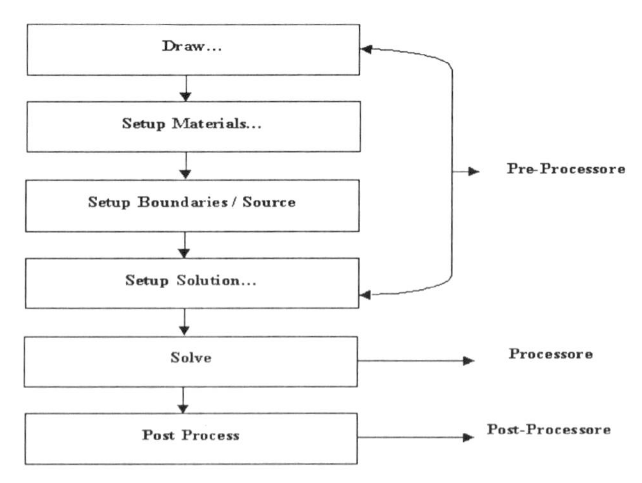

Figure 3: Design flow

In the next section the required steps to realise a design with this tool will be presented.

Draw

It allows to create structure geometry using a three-dimensional graphic interface. This is the first step to follow for successfully developing a design.

When a project has to be realised, the following rules must be fixed:

- don't use many vertices on curved shapes where it isn't necessary. Often 8 or 12 facets on a circle are more than adequate. More vertices increased the computational effort without particular advantage on solution accuracy;
- avoid narrow gaps between parallel faces of objects. These can lead to an explosive growth in the number of mesh elements;
- use smaller two-dimensional objects, called *dummy object*, to thicken the mesh on a particular interest region or by manual mesh refinement;
- keep the boundary of the problem region sufficiently far away so that it doesn't significantly influence the field in the vicinity of the model objects. Keep it sufficiently close so that a large volume of space with very low field values isn't include. As a rule of thumb, the region should be between 5 to 10 times the linear dimension of the modelled objects;
- exploit symmetry in the design if possible. For a given solution accuracy, a single symmetry plane reduced the computational effort by a factor of 4, two planes of 16, and so on.

Setup Materials

The comprehensive materials database contains permittivity, permeability, electric, and magnetic loss tangents for common substances. Users may include homogeneous, inhomogeneous, anisotropic, conductive, resistive, and semiconductor materials in the simulation.

Setup Boundaries / Source

It allows to assign boundaries condition and signals or sources on the different objects belonging the structure.

Boundaries

Boundary conditions are applied on objects surface present on the model to be analysed. By default, the electric field is assumed to be normal to all surfaces exposed to the background (region that surrounds the geometric model and fills any space not occupied by an object) representing the case in which the entire structure is surrounded by perfectly conducting walls. This tool has different types of boundaries condition such as:

perfect-E boundary: sets the component of E-field to be normal to the surface selected, it allows users to model perfectly conducting surfaces;

perfect-H boundary: forces the tangential component of the H-field to be the same on both sides of the boundary;

finite conductivity: define all object surfaces that represent imperfect conductors;

impedance: models a surface as a resistive surface which thickness is bigger than skin effect depth. The behaviour of the field at the surface and the losses generated by the currents flowing inside the resistor are computed using analytical formulas;

radiation: simulates open problems, such as antenna problem, that allow waves to radiate infinitely far into space. The system absorbs the wave at the radiation boundary, essentially ballooning the boundary infinitely far away from the structure. A radiation surface doesn't have to be spherical, however, it must be exposed to the background and located at least ¼ of a wavelength away from the radiating sources;

symmetry planes: allows to divide the geometric structure. Reducing the size or complexity of the problems helps speed up the solution time. If the symmetry is such that the electric field (E) is normal to the symmetry plane, it is necessary to use a Perfect-E. If the electric field (E) is tangential to the symmetry plane, a Perfect-H must be used.

Sources

Sources applied on geometric model are:

Port: by default, the interface between the background and all objects is assumed to be a perfect electrical conductor (the interface is a Perfect-E boundary), so no energy is able to enter and exit the structure. Ports represent the portion of the structure through which signals enter or exit. HFSS generates a solution by exciting each port individually. Port 1 is excited by a signal of 1 watt and the other ports are set to zero watts. After a solution is generated, port 2 is set to 1 watt and the other ports are set to zero watts and so on. In solving for the S-parameters, the software assumes that the structure is excited by the natural field pattern (modes) associated with these cross-sections. The two-dimensional field solution generated for each port serves as boundary conditions at those ports. The final field solution computed must match the two-dimensional field pattern at each port. When a port is defined, is necessary to take different things into consideration:

- only surfaces that are exposed to non-existent objects, such as the background or objects defined as perfect conductors, can be defined as ports;
- a port must lie in a single plane and ports that bend are not allowed;
- when designing a problem containing ferrite materials and ports, don't arrange the port so that it touches the ferrite material.

Incident Wave: allows to define the electric field (E) as a wave that both propagates in one direction and is uniform in those directions perpendicular to its direction of propagation.

Voltage Drop: defines the voltage and direction of the electric field on a surface. This source is specified in a peak sense.

Current: defines the amplitude and direction of the current flow through a surface.

Setup Solution

It allows to select the type of solution to be performed. Three types of solution are available:

adaptive solution: the system selectively refines the finite element mesh. Based on the current finite element solution, the system estimates the regions of the problem domain where the exact solution has strong error. Tetrahedra in these regions are refined;

non-adaptive solution: the mesh is fixed by user and the solution is generate on it;

frequency sweep: generates a solution across a range of frequencies. A *fast* or a *discrete sweep* must be chosen. Discrete sweep must be selected if only a few points are necessary to accurately represent the results. However, if the model resonate or change operation in the frequency band, a fast sweep must be chosen to obtain an accurate representation of the behaviour near the resonance.

Post—Processor

It shows the design results in a graphic way. In particular:

- it realises diagram animation of the near and far field inside the structure;
- it views matrices computed for S-parameters, port impedances, propagation constants, admittances and impedances during each solution;
- it plots several parameters evaluated in time or frequency function.

HSB2 Technology

High-Speed Bipolar (HSB2) technology is used to realise high-frequency application bipolar transistor, particularly in radio-frequency (RF) field [1].

The development complexity of this technology can be explained as follow:

- 14-17 photolithographies (minimum geometry 1 μm);
- 8 ions implantation;
- 2 polysilicon levels;
- 2-3 metal levels.

The section of a NPN BJT (HSB2P46 C) is shown in Figure 4.

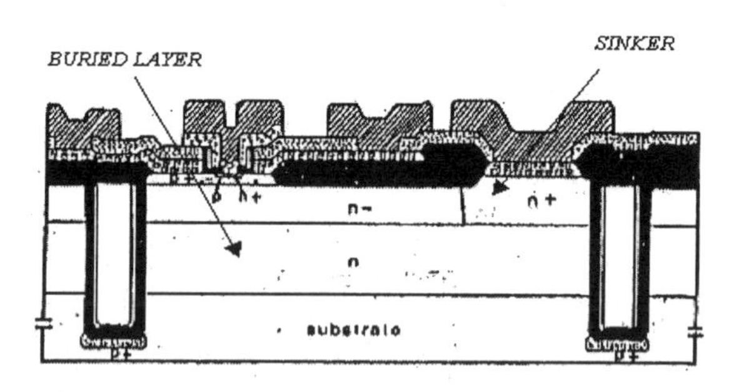

Figure 4: Section representation of a BJT (npn)

The collector region, called Buried-Layer (N^+), has high conducibility to obtain a low collector resistance value. Regarding the metallization contact, the layer appears on surface through the Sinker, which represents a layer of n-type diffusion.

The isolation, by one device to another, is obtained through trenches constituted by depths holes with walls in silicon oxide and silicon amorphous to its inside. Such structures are shown in Figure 5.

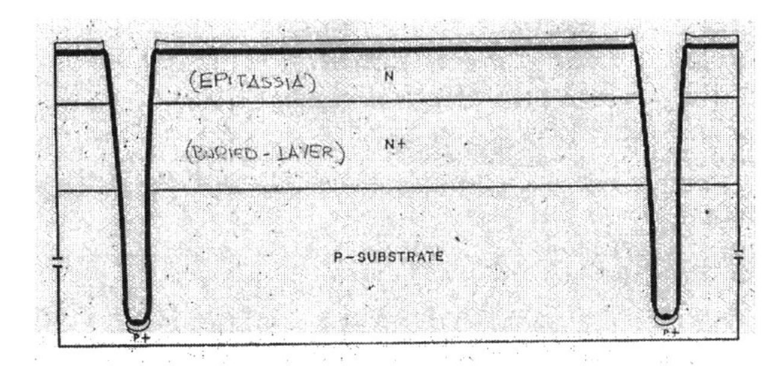

Figure 5: Trenches isolation

The base and collector contacts are obtained by polysilicon and the frequency cut is around 20 GHz. These represent the fundamental characteristics of this technology for the radio-frequency.

Application and Results

Microstrips in metal 3

In order to estimate the performance results of a tool it is necessary to compare them with accuracy measurements carried out in the laboratory.

Due to the availability of an experimental setup, a microstrip with HSB2 (High-Speed Bipolar) technology has been implemented using the HFSS tool simulator [2].

The microstrip is in metal3 with 1mm length, 10 μm width and 1.5 μm thickness; its structures is reported in Figure 6.

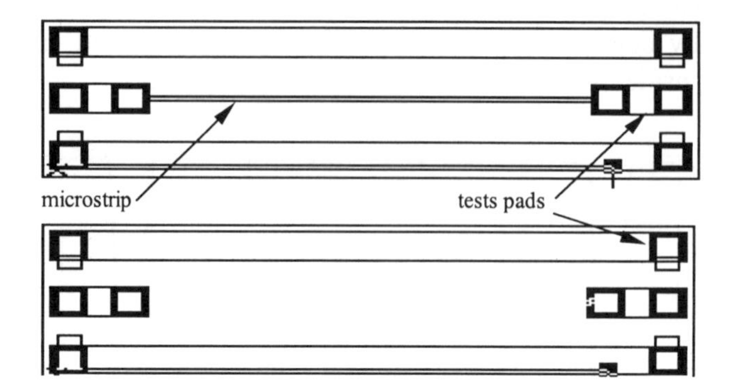

Figure 6: Layout of the microstrip with deembedding structure

The measurements on such microstrip carried out in the R& D Electronic Laboratory, using deembedding structure to avoid the parasitic effects present in the layout of the microstrip, are reported in the Figures 7 and 8.

Such measurements, are shown and explained in a ST-report [2].

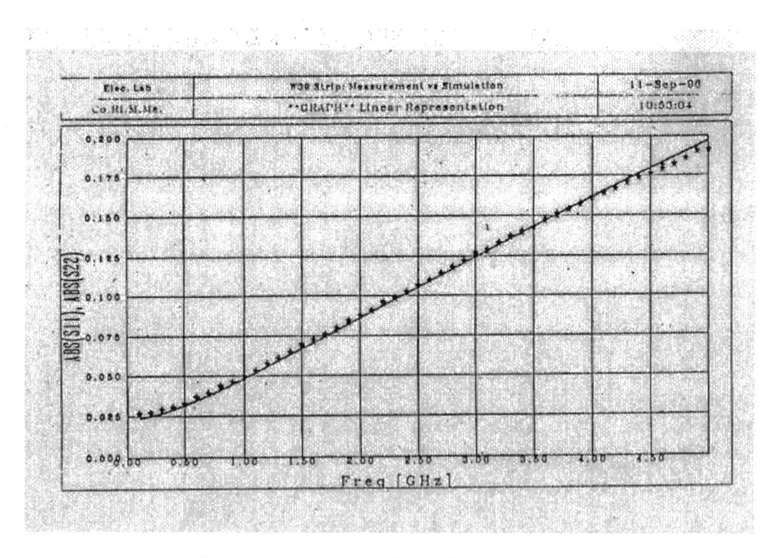

Figure 7a: Magnitude and Phase of S_{11}

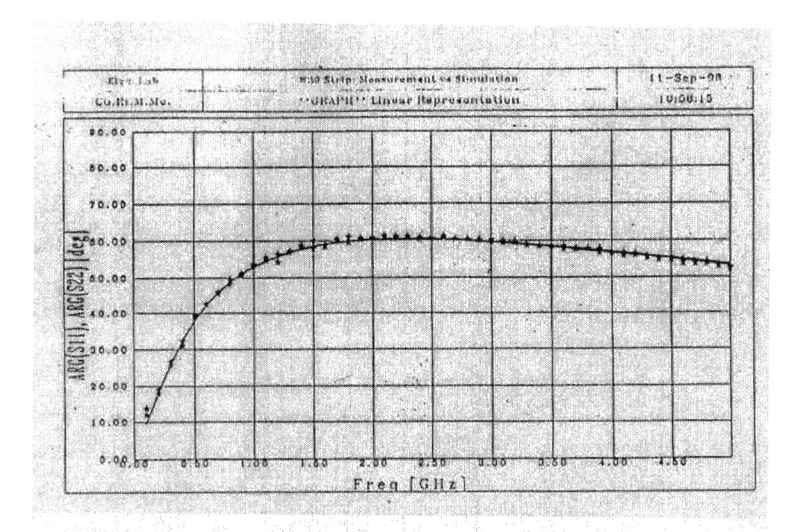

Figure 8b: Magnitude and Phase of S_{22}

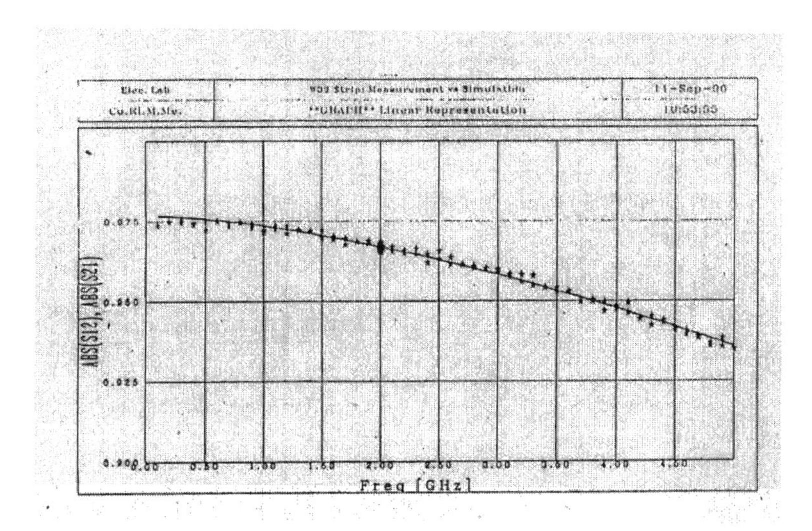

Figure 8a: Magnitude and Phase of S_{12}

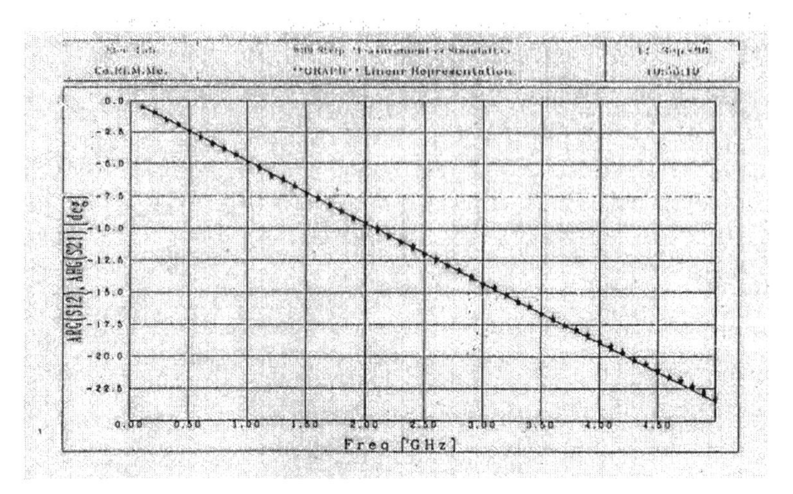

Figure 8b: Magnitude and Phase of S_{21}

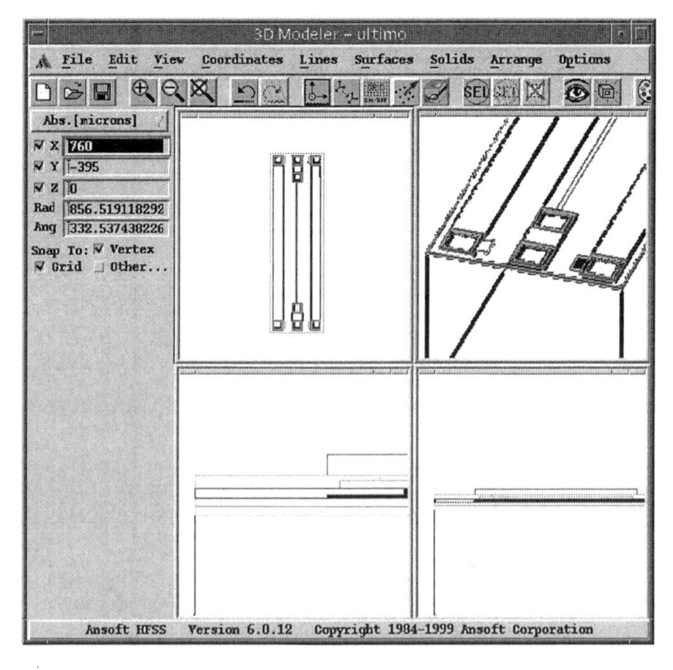

Figure 9: Three-dimensional structure of the microstrip with HFSS

Microstrip design with HFSS

The three-dimensional model of the microstrip has been realised considering a multi-layer structure that coincides, with the implementation processes of HSB2 technology. For each layer the material properties and the relative thickness have been defined.

In Figure 9 the microstrip implementation with HFSS tool is shown.

Results

In Figure 10 a three-dimensional microstrip symmetry is shown. In these pictures both the Electric-Field lines and their intensity around the microstrip are represented.

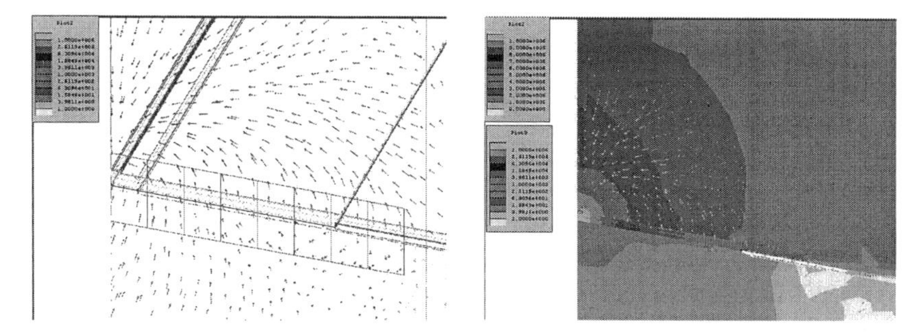

Figure 10: Electric field and intensity lines

During the project development, several difficulties, not all overcome, have been found to define the structure ports, according to the following rule.

"Ports must be enough great so that its sides don't influence the field lines of the implemented structure and at the same time they must be small in such way that an ulterior propagation mode inside the microstrip doesn't exist".

The results, obtained by the simulator and reported in Figures 11 and 12 show the variation of the S-parameters for different values of the ports dimension.

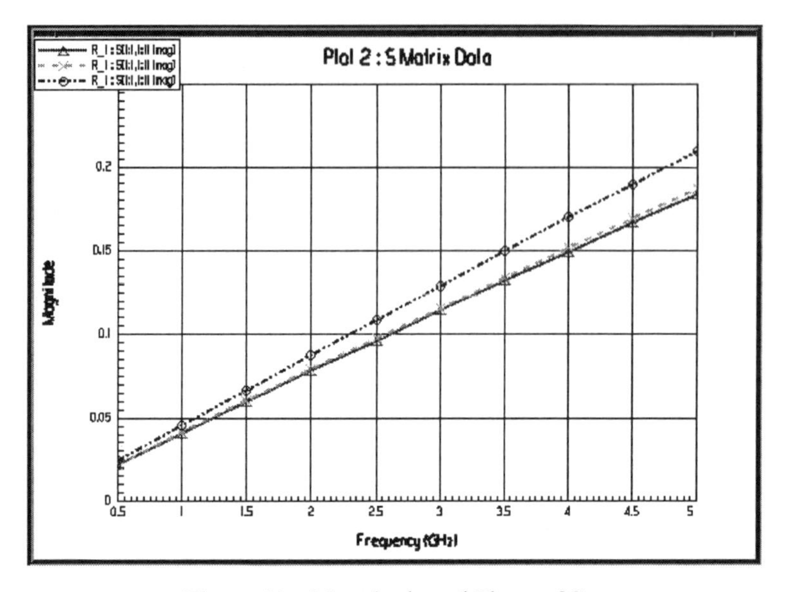

Figure 11a: Magnitude and Phase of S_{11}

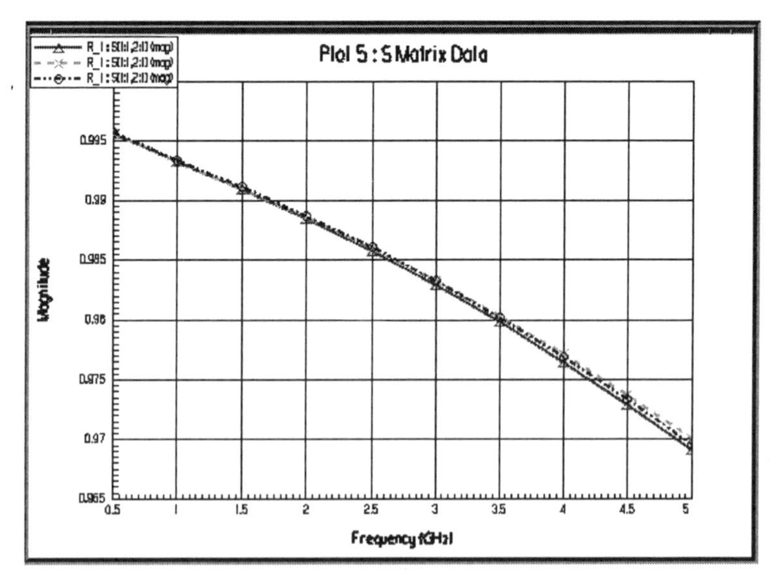

Figure 11b: Magnitude and Phase of S_{22}

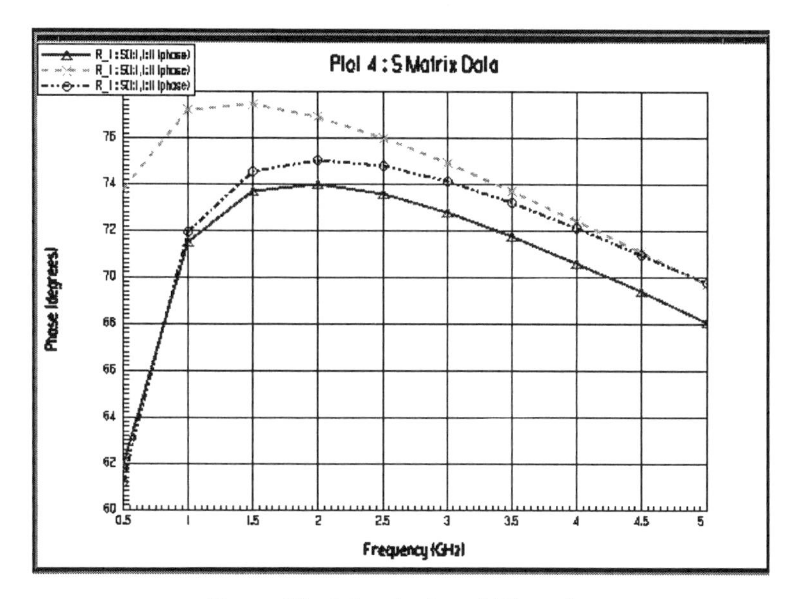

Figure 12a: Magnitude and Phase S_{12}

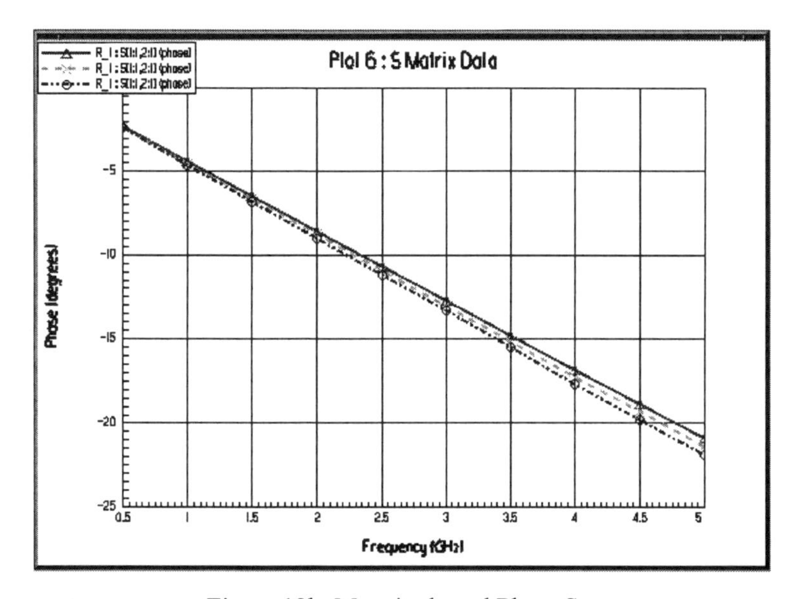

Figure 12b: Magnitude and Phase S_{21}

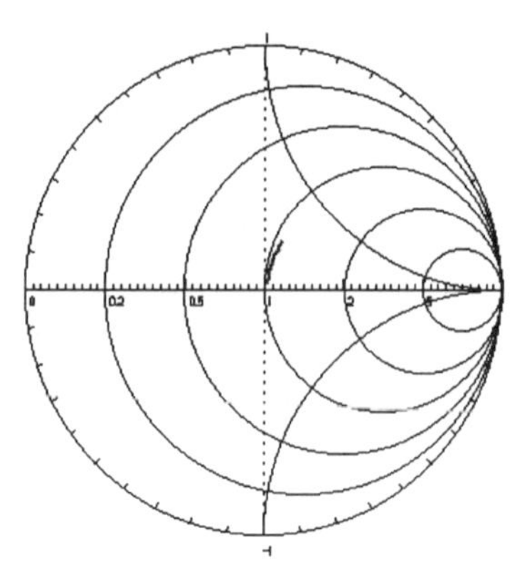

Figure 13: Smith Chart

The result in Figure 13 shows that the passive component implemented with HFSS tool has an inductive behaviour.

The simulations have been run for different ports dimension. Taken as reference the rectangle centred on the microstrip with dimension W=180mm, L=250mm (line with symbol x), the same has been increased by 20% (line with symbol Δ) and by 40% (line with symbol o).

Conclusions

Although the passive component characterisation points out that the model used is qualitatively valid, there is a numeric disagreement between measured data in the laboratory and simulated data with this tool. This is due to both to the dependence of S-parameters on the uncertainty in the ports dimension and to a different definition of the boundary conditions in a multilayer structure.

References

[1] Pinto, A., *Tecnologia HSB2.*Co.Ri.M.Me.Catania, 1996.
[2] Privitera, G.- La Rosa, G., *Caratterizzazione della μstrip line in metal 3 di larghezza 10 μm in tecnologia HSB2P, RT172.* Laboratorio Elettronico R&D ST-Microelectronics. Catania, 1996.

Integrated Passive Components and Power Amplifier Elementary Stadiums in Integrated HF Technology

Gianluca Catrini

STMicroelectronics S.r.l.

Stradale Primosole 50, 95100 Catania - Italy

E-mail: gianluca.catrini@st.com

Introduction

Recently, interest in on-chip spiral inductors and transformers has surged whit the growing demand for radio frequency integrated circuits (RF IC's).

For silicon-based RF IC's, the inductor quality factor (Q) degrades at high frequency due to energy dissipation in the semiconducting substrate; furthermore, they occupy substantial chip area.

Fine-Line CMOS technology easily provides high frequency active devices for use in RF applications (e.g., 800 Mhz-2.4Ghz), but high quality passive components (e.g., inductors) present serious challenges to integration as exemplified by several recently reported CMOS RF low-noise amplifier (LNA) designs [3].

Although significant progress toward the integration of high quality inductors including many innovative structures and design techniques, practical planar monolithic inductors have achieved only moderate performances owing to resistive losses in the metal traces and in the underlying substrate.

For these reasons the understanding and optimisation of passive integrated components have received intense attention.

In this job, the passive components development (inductors and transformers) for integrated HF technology, is treated with electromagnetic simulation tools by the AnSoft Corporation.

AnSoft Electromagnetic Simulators

The complexity and speed of today's electronic system demand consideration of electrical issues such as signal integrity and EMI/EMC to prove design concepts prior to prototyping. Ignoring these effects during the design phase could prove devastating.

Accurate signal integrity analysis of high density and high speed interconnections is imperative in package design. As die dimensions and clock frequencies increase, signal wavelengths become comparable to wire lengths and require interconnects to be modelled using full-wave field solvers.

AnSoft Corporation has developed a software package suitable for a complete design flow. With the Serenade Environment Design, it's compatible to other CAD environment.

The Serenade Design Environment is an integrated software suite for RF and microwave design that delivers all high-frequency design technologies: system, circuit, electromagnetic simulation, synthesis, and physical design. Engineers can access schematic capture, high-frequency circuit, electromagnetic, and system simulators, layout, IC package modelling, and links to third-party tools. Handshaking between tools allow engineers to focus on critical components at any stage of engineering development. Each simulator can be used as a powerful, stand-alone tool or in concert with others for end-to-end high-frequency design. Serenade offers flexibility plus integration without rigid hierarchy.

HFSS

Ansoft HFSS is an interactive software package that computes s-parameters and full-wave fields for arbitrarily shaped 3D passive structures. It offers an intuitive interface to simplify design entry, a field-solving engine with accuracy-driven adaptive solutions.

To generate the solution of electromagnetic fields, HFSS brings the power of the finite element method (FEM). The software also includes post-processing commands to analyse the electromagnetic behaviour of the structure in detail.

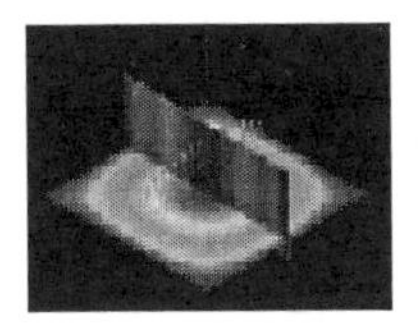 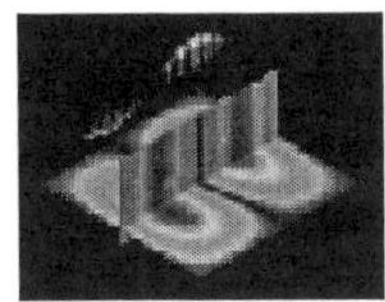 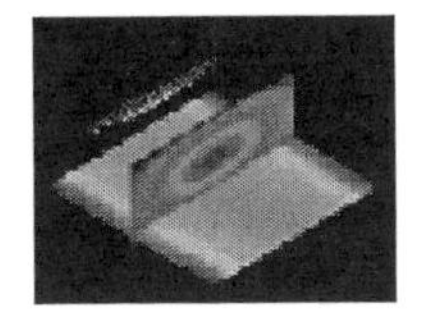 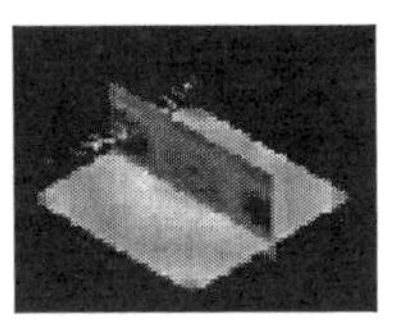

Figure 1: Representation of fields in HFSS

The Finite Element Method (FEM)

In order to generate an electromagnetic field solution, Ansoft HFSS employs the finite element method.

In general, the finite element method divides the full problem space into thousands of smaller regions and represents the field in each sub-region (said element) with a local function.

The geometric model is automatically divided into a large number of tetrahedrons, where a single tetrahedron is basically a four-sided pyramid. This collection of tetrahedrons is referred as the finite element mesh.

Implementation

To calculate the S-matrix associated with a structure, the system does:

- divides the structure into a finite element mesh;
- computes the waves on each port of the structure that are supported by a transmission line having the same cross-section as the port;
- computes the full electromagnetic field pattern inside the structure, assuming that each of the ports are excited by one of the waves;
- computes the generalised S-matrix from the amount of reflection and transmission that occurs.

The final result is an S-matrix that allows the magnitude of transmitted and reflected signals to be computed directly from a given set of input signals, reducing the full three-dimensional electromagnetic behaviour of a structure to a set of high frequency circuit values.

Size of mesh versus accuracy

There is a trade-off between the size of the mesh, the desired level accuracy and the amount of available computing resources.

The accuracy of the solution depends on how small each of the individual elements (tetrahedrons) is. Solutions based on meshes using thousands of elements are more accurate than solutions based on coarse meshes using relatively few elements. To generate a precise description of a field quantity, each element must occupy a region that is small enough for field to be adequately interpolated from the nodal values.

However, generating a field solution involves inverting a matrix with approximately as many elements as there are tetrahedron nodes. For meshes with a large number of elements, such an inversion requires a significant amount of computing power and memory. Therefore, is desirable to use a mesh fine enough to obtain an accuracy field solution but not so fine that it overwhelms the available computer memory and processing power.

To produce the optimal mesh, HFSS uses an iterative process in which the mesh is an automatically refined in critical region. First, is generated a solution based on a coarse initial mesh. Then, it refines the mesh in areas of high error density and generates a new solution. When selected parameters converge to within a desired limit, the system breaks out of the loop.

BCD Technology for Smart Power ICs

The name BCD has been created in the mid-eighties to classify that family of mixed processes which allow to integrate into a single chip Bipolar, CMOS and DMOS transistors, forming a new power IC class called SMART POWER ICs [6].

Up till the mid-eighties Power ICs were realised using pure bipolar technologies previously developed together with their main application field, the audio amplifier market segment.

For the analog function required in these applications, the bipolar transistors were the best choice due to their amplification and matching properties.

The other major application field was the motor control one where the needed logic functions where implemented by bipolar I^2L structures. With the growing requests of logic functions the bipolar I^2L approach resulted limited due to their design complexity, their high power consumption and their limited lithography.

As these limits were not present in the CMOS components, at least at low frequency operation, to increase the smartness in the Power ICs it appeared evident that the only way was to make available CMOS in the place of I^2L.

During the Bipolar IC evolution the power that had to be supplied to the load increased, too. This increase reached the limit fixed by the allowed power dissipation and by the admitted consumption.

The DMOS power transistor permitted to overcome all these limitations because it has no driving DC current requirement and it efficiently works in fast switching conditions.

The previous topics have been realised with the idea to merge the discrete DMOS technology, rising in the first eighties, with the junction-isolation bipolar technology used for many years to construct Power Integrated Circuits.

The realisation of this idea led to the birth of the Multipower-BCD. Several different technological approaches to smart power have been explored, but all such technologies share the same basic concept of merging different structures into the same chip, taking advantage of the similarities in processing techniques.

Integrated power structure

The most common power device used in a Smart Power Integrated Circuit is the DMOS transistor for the well-known advantages that it offers compared to bipolar transistors (Table 1).

Several kinds of new power devices have been developed to compete with the performances of DMOS in high voltage and high current applications, but all these approaches are only aimed at market niches. Besides, as these are SCRs or thyristor-based structures, to overcome the problems coming from the current injection into the substrate, that can be a limiting factor to make them suitable as power elements in an integrated circuit.

Vertical & horizontal structures

As far as the geometry of power device integration is concerned, the positioning of the drain/collector contact on the surface or the rear of the chip is determined not only by the technology but also by the needs of the application.

MOS	BIPOLAR
• Majority carrier device. No charge storage effects; high switching speed; drift current.	• *Minority carrier device.* Charge store in the base and collector; low switching speed; diffusion current.
• Voltage driven. High input impedance; no dc driving current consumption; simple drive circuitry.	• *Current driven.* Low input impedance; dc driving current consumption; complex drive circuitry.
• Negative temperature coefficient of drain current for V_{GS} slightly higher than V_{TH}. No thermal runaway; No second breakdown limitation.	• *Positive temperature coefficient of collector current.* Thermal runaway effect; susceptible to second breakdown.
• *Low transconductance.*	• *High transconductance.*
• *Drain current is proportional to channel perimeter over length ratio.*	• *Collector current is proportional to emitter area.*

Table 1: Comparison of MOS and bipolar power transistor

In a vertical structure the current flows through the device from the top to the bottom, across the substrate and the die attach area to the package itself as it is shown in Figure 2. This scheme, with the substrate above ground potential, is like that of discrete power devices with the addition of an isolated structure containing the control circuitry. The vertical structure allows very high current densities but there is the limitation that in one chip it is only possible to integrate one power transistor, or several with the collectors or drains shared.

In a horizontal structure the current enters and leaves the chip through the upper surface of the chip. In this case, too, an isolated structure can be added that contains the control circuitry. This structure is derived from those adopted in standard power integrated circuits where the substrate is simply a mechanical support and a heat conductor. In the horizontal structure several variations can be distinguished on the basis of the current and voltage required.

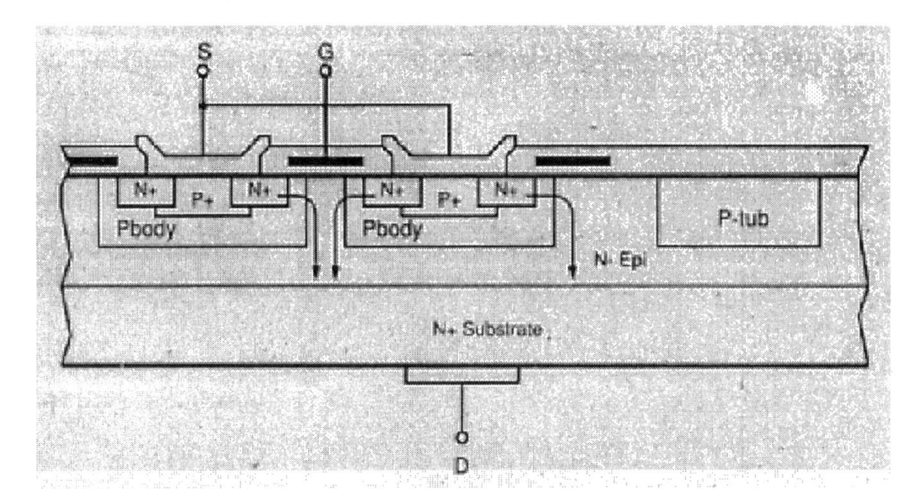

Figure 2: Current flow in a vertical structure

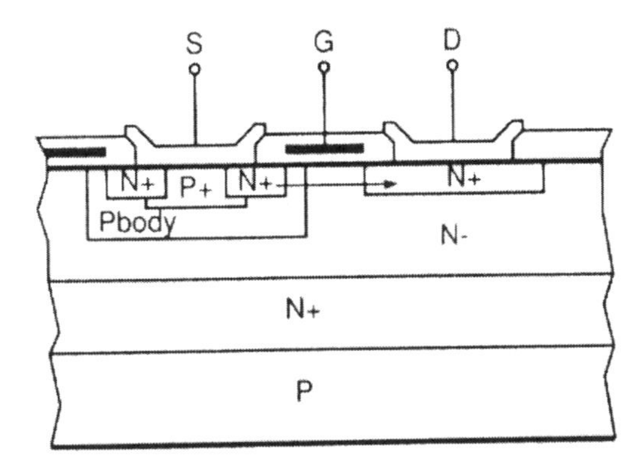

Figure 3: Current flow in horizontal structures

BCD6 technology

The BCD process evolution has developed not only in the voltage direction, but also towards the direction of the minimum lithography that is today scaled down to 0.5 μm.

In fact another important advantage of the BCD is the dramatic increase of complexity induced in the Power ICs. This has been possible because in the BCD technology the power component in a MOS transistor where the current density depends on the geometrical ratio W/L (channel width/channel length) which can be improved simply by reducing the line width. Differently, in the power bipolar components the current density depends on the emitter area and therefore progress in micro-lithography does not bring appreciable improvement in density.

The BCD6 technology is a natural evolution of previously and actually it's in development.

The mind layers of this new technology are shown in the following Table:

	Thickness µm	Resistivity ρ	Doping cm-3
Substrate P+	390	20 mΩ·cm	-
Substrate P-	7	10-20 Ω·cm	-
Well (triple n-well)	3	-	10^{16}
Field Oxide (active region and	0.5	-	-
Metal1 (tungsten)	0.6	-	-
Oxide	0.9	-	-
Metal 2,3,4 (aluminium or copper)	0.6	-	-
Metal5 (aluminium or copper)	2.5	-	-
Passivity			
- **No doped oxide**	1	-	-
- **Doped oxide**	0.5	-	-
- **Silicon nitride**	0.7	-	-

Table 2: The mind layers of new technology

Passive Elements Development

The continuous evolution of silicon technologies is introducing some innovations in the RF design of transceiver for wireless communications, in particular on-chip spiral inductors and transformers. Currently, technology scaling allows CMOS and BiCMOS processes to operate at RF and great effort is underway to obtain a monolithic solution that meets mobile telecommunication standard specifications. The goal of a COMS or BiCMOS single chip with low power consumption and reduced fabrication cost can be met only if these technologies succeed to integrate good quality passive components. Therefore, inductors need to be realised on a silicon substrate along with all of the other devices in a single chip. The need for high Q integrated inductors in RF ICs is increasing to be used in fully integrated VCO (voltage controlled oscillator). These devices are also typically used in low voltage/low power designs to realise narrow-band impedance matching, tuned loads (resonant tanks), low noise degeneration and feedback and linear filters with high dynamic range.

Integrated inductors on GaAs have been successfully realised with Q values on the order of tens. The lower substrate resistivity of silicon especially for CMOS processes reduces the Q values due to the increased substrate losses. For these reasons it is important, in CAD applications, to model these effects using an accurate lumped-element circuit deduced from experimental or numerical characterisation [1].

Acquired the fundamental notions relative to the technology and simulator, we started the development of passive components for integrated RF technology, with particular care to inductors and transformers to be applied in Power Amplifier.

Integrated inductors

In Figure 4 a simplified cross section of an inductor in BCD6 technology is shown.

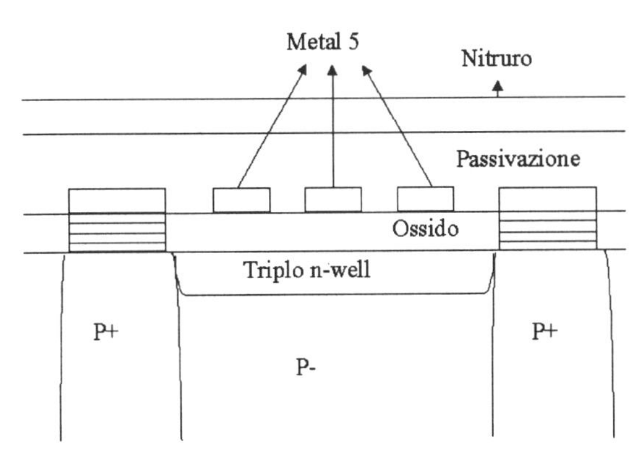

Figure 4: Cross section of an inductor in BCD6 technology

As seen, the technology is composed of five layers of metals. The inductor is realised on the last layer, because this has a bigger thickness and so we can reduce the resistivity loss due to the material. The upper part of the substrate is a 7μm thick P^- layer over a P^+ layer 390μm thick. A 3μm n-triple-well has been drawn under the transformer in order to increase the isolation from the substrate.

In Figure 5 an inductor realisation in HFSS environment is shown.

The Figure shows the top level of metal (the less resistive), used to realise the spiral, and a fourth metal level is used to provide access to the second port of the device.

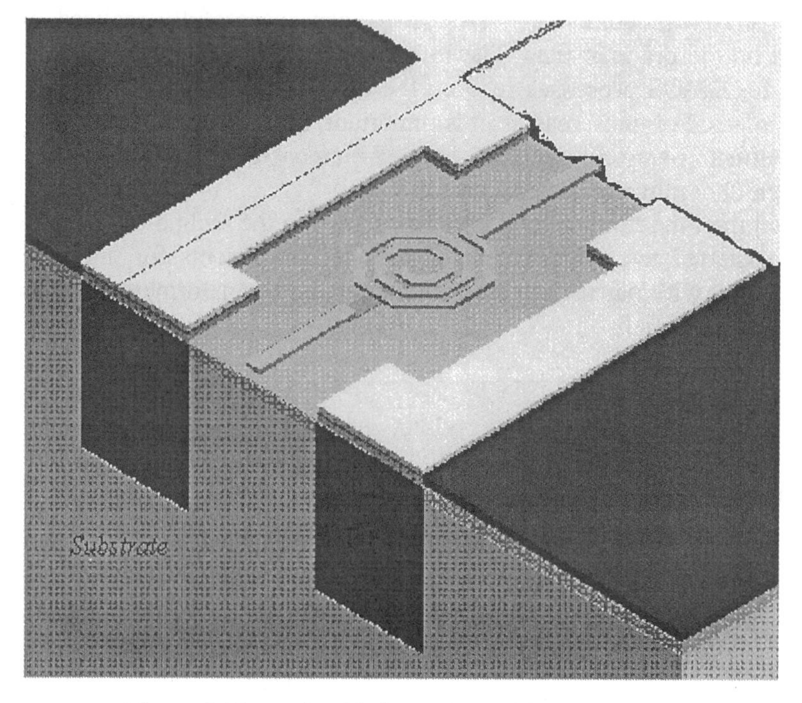

Figure 5: Example of inductor in HFSS environment

Since these devices are designed for low cost applications, the technology used should be completely standard. Therefore, a distinction must be made between design and process parameters.

Design parameters (area, shape, metal strip width, distance between metals and number of turns) can be manipulated by the designer to optimise the inductor's features; process parameters (substrate resistivity and metal line conductivity) are set by the technology. This constraint is critical because a silicon substrate is intrinsically characterised by high losses; moreover a P^- layer, whose thickness is set to 7μm, has been put under the inductor in order to decrease the losses. Moreover, metal resistivity contributes to a further degradation of the Q of these devices. The substrate contact, around triple n well, is made with P+ type contacts.

The entire structure is closed in a box, with vacuum characteristic, in order to have a closed problem. In this way the base of the structure, that touches the background, represents the ground plane.

Simulation set-up

Five different electromagnetic simulations have been carried out, by varying the operating frequency in the range 1.5-3.5 GHz with a step of 500 MHz.

Using the following set-up has simulated the scattering parameters associated to the inductor:

- Physic characteristics: conductors used to realise the inductor (metal5) are Aluminium (Al) and Copper (Cu).
- Geometric characteristics are:
  - metal5 thickness: 2.5 and 5 μm;
  - metal4 thickness: 0.6 μm;
  - width: 5 and 10 μm;
  - spacing between the coils: 5 and 10 μm;
  - intern radius: 10 and 30 μm;
  - number of turns: 2.5.

Varying these parameters the mesh and then the field into the structure and on surface of the structure changes.

In Figure 6 an example of mesh achieved with finite element method is shown.

The mesh is more refined in the regions major interest (input and output of inductor). This increase in complexity (thus computation time) has forced to obtain a more accurate solution.

With HFSS post processor, from S-matrix it's possible to compute Y-matrix and Z-matrix for each set of parameters.

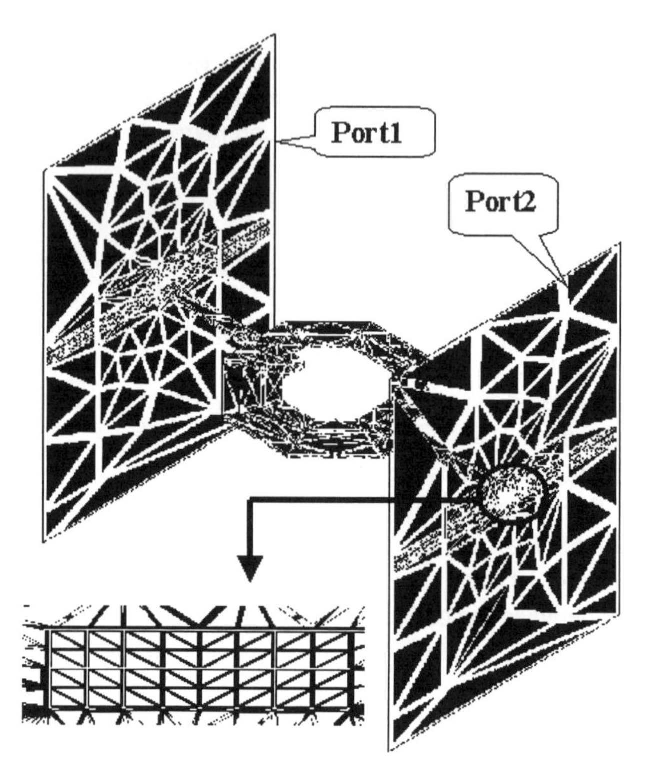

Figure 6: Subdivision for the structure in finite elements

Definition and measurements of quality factors

There are at least two widely used definitions for Q factor.

The first and probably the most fundamental definition is based on the maximum energy storage and average power dissipation which for simplicity, is referred as Q_E [2]:

$$Q_E = \frac{\omega \cdot W_{max}}{P_{diss}}$$

ω is the radian frequency and Wmax is the maximum total electrical and magnetic energy stored in the system.

Unfortunately an accurate estimation of QE is difficult.

The most widely used Q definition, which is referred as Q_C, is the ratio of the negative imaginary part of y11 and the real part of y11.

The y11 data are obtained by converting measured S-parameters of inductors.

A common equivalent circuit to model an integrated inductor in silicon IC processes is shown in Figure 7.

The y11 data are utilised for the quality factor computations because they represent the admittance seen looking at port 1 while port 2 is shorted to ground. This is a common configuration in which the inductors are used in amplifiers and oscillators.

Using simple network theory, it can be shown that it results

$$Q_C = -\left[\frac{\mathrm{Im}(y11)}{\mathrm{Re}(y11)}\right] = \frac{2\cdot\omega\left(\left|\overline{W_m}\right| - \left|\overline{W_e}\right|\right)}{P_{diss}}$$

where $\left|\overline{W_m}\right|$ e $\left|\overline{W_e}\right|$ are the average stored magnetic and electrical energies in the system. When port 2 (P2) of equivalent circuit model in Figure 7 is shorted to ground, $\left|\overline{W_m}\right|$ e $\left|\overline{W_e}\right|$ are the energies stored respectively in the inductor L_S and capacitor C_{OX_EXT}.

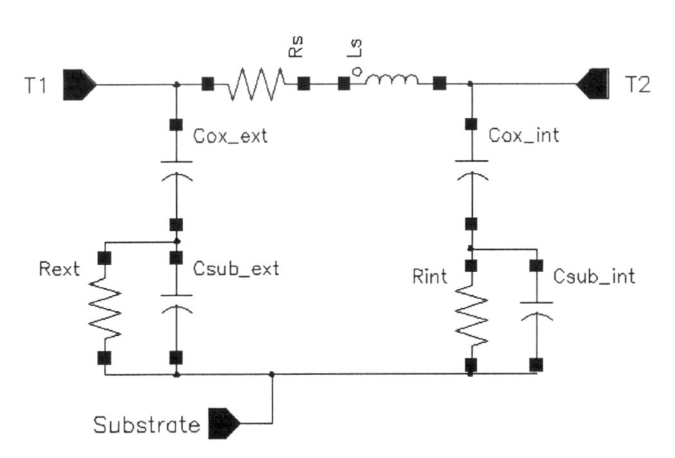

Figure 7: Equivalent circuit used to model the integrated inductors

This Q definition involves the difference between the average stored magnetic and electrical energies rather than the maximum total energy storage.

When the average magnetic energy storage is much greater than the electrical storage, this ratio approaches Q_E. For the general case as well as for silicon integrated inductors with typically large shunt capacitance to the substrate, thus, significant electrical energy storage, Q_C can deviate from Q_E by a large amount.

	Operative Frequency				
	1.5 GHz	**2 GHz**	**2.5 GHz**	**3 GHz**	**3.5 GHz**
L_S	0.434 nH	0.423 nH	0.426 nH	0.443 nH	0.474 nH
R_S	0.344 Ω	0.308 Ω	0.388 Ω	0.391 Ω	0.403 Ω
C_{ox_ext}	13.81 pF	0.853 pF	68.89 fF	66.56 pF	53.286 fF
C_{ox_int}	3.888 pF	5.838 pF	2.503 fF	46.05 pF	11.23 fF
R_{int}	191 Ω	374 Ω	230 Ω	198 Ω	226 Ω
R_{ext}	192 Ω	345 Ω	69 Ω	118 Ω	125 Ω
Number of turns : 2.5					
Conductor type: Aluminium					
Thickness, width, spacing, internal radius: 2.5μ, 5μ, 5μ, 10μ					

Table 3: Model parameter values

	Operative Frequency				
	1.5 GHz	**2 GHz**	**2.5 GHz**	**3 GHz**	**3.5 GHz**
L_S	0.385 nH	0.381 nH	0.378 nH	0.376 nH	0.376 nH
R_S	0.252 Ω	0.272 Ω	0.28 Ω	0.27 Ω	0.257 Ω
C_{ox_ext}	32.43 fF	20.33 fF	62.47 fF	14.91 pF	24.89 fF
C_{ox_int}	15.72 fF	27.62 fF	17.22 fF	14.96 pF	22.03 fF
R_{int}	223 Ω	23 Ω	230 Ω	134 Ω	148 Ω
R_{ext}	148 Ω	459 Ω	69 Ω	138 Ω	182 Ω
Number of turns : 2.5					
Conductor type: Copper					
Thickness, width, spacing, internal radius: 5μ, 5μ, 5μ, 10μ					

Table 4: Model parameter values

Results

In Table 3 and 4 are reported the model parameter values computed in correspondence to the physic and geometric characteristics and for a range of frequencies.

In Figure 8 the factor quality (Q_C) versus frequency are reported.

In particular the quality factor for aluminium (thickness 2.5 μm) and copper (thickness 5 μm) are plotted.

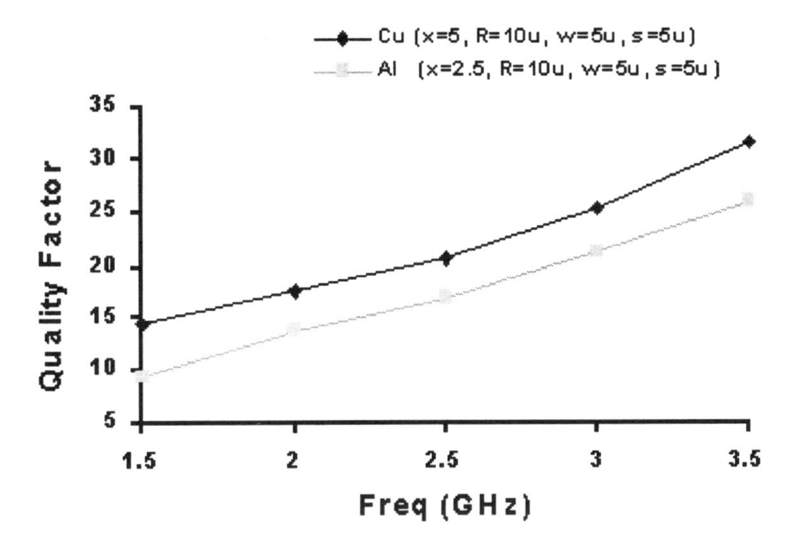

Figure 8: Q_C versus frequency

x = thickness; R = internal radius; w = width; s = spacing

Analysis of results

With reference to the simulation results, we can note that the inductance values are in agreement with the expectations. But the Q values are bigger than the expectations, particularly increasing the frequency.

It can occur because increasing the frequency, the electrical energy storage increases, and the difference between the average stored magnetic and electrical energy decreases, which in turn increases, the difference between Q_C and Q_E. As a matter of fact, since the self-resonant frequency of an inductor occurs near a frequency where the difference between the average stored magnetic and electrical energy is zero, or where the imaginary part of y11 is equal to zero. Q factors, extracted using Im(y11)/Re(y11), becomes zero near the self-resonant frequency. Of course this result is physically unreasonable. The quality factor should not be zero at the self-resonant frequency.

Lastly, it was stated earlier that accurately estimating Q_E is difficult. This is due to lack of modelling of others parasitic as well as inductive elements. Related difficulties include the distributed nature of model elements and their frequency dependence.

This may mean that the equivalent circuit model at best is not very accurate. These problems are exacerbated by the fact that the model parameters are extracted using y-parameters that depend on the difference between the average magnetic and electrical energies rather than on the maximum total energy storage.

Integrated transformers

In Figure 9 the transformer drawing in HFSS environment is shown.

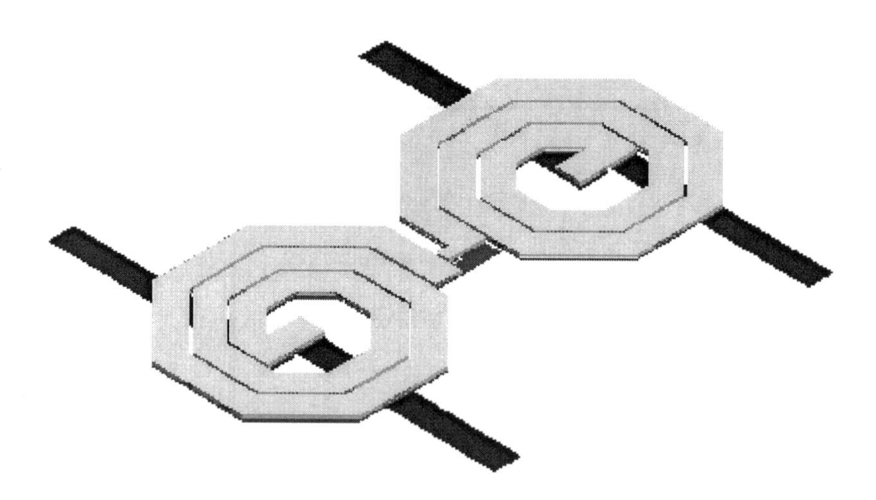

Figure 9: Transformer in HFSS environment

The realisation of this device is in progress. The base idea is to realise a symmetric transformer. The primary, as well as the secondary, is made both in metal4 and metal5. In this way we try to balance the resistive losses due to the different thickness between metal4 and metal5. The terminals are in metal3.

The next step is to use some simulation tools, as ADS (Advanced Design Systems), to extract a model from scattering parameters obtained with HFSS.

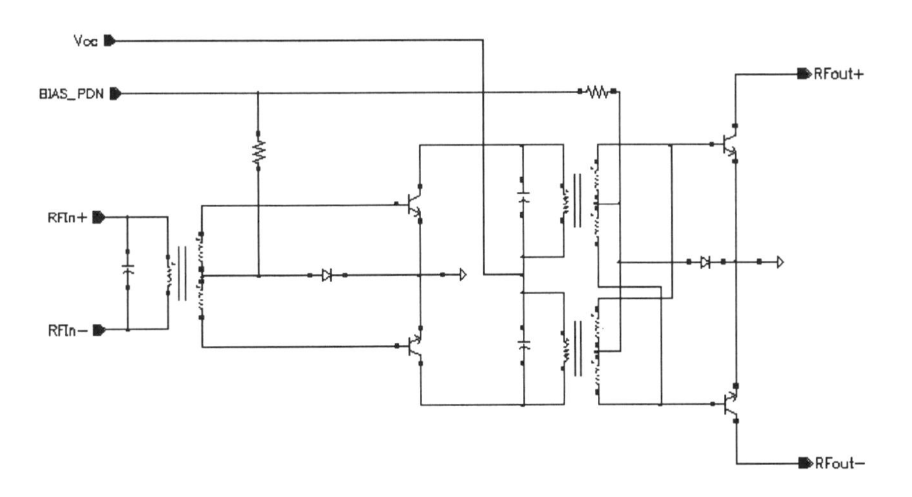

Figure 10: Power amplifier with integrated transformer

Conclusions

A possible application concerning the realisation of power amplifier with integrated circuits with transformers reported in Figure 10. The following tasks can be realise introducing in power amplifier integrated transformers:

- input port matching with differential and single ended (with a port to ground) signals;
- matching for intermediate stadiums;
- no block dc capacitance.

Such devices can be competitive thanks due to the reduced number of external components. Moreover power amplifier for RF applications, with reduced cost and high efficiency, are really necessary for today mobile communications.

References

[1] Microwave JOURNAL, Euro-Global Edition, *CAD and Measurement*, VOL.42, NO. 8, August 1999.

[2] Kenneth O., *Estimation method for quality factors of inductors fabricated in silicon integrated circuit process technologies*, IEEE JOURNAL of solid-state Circuits, VOL.33, NO.8, August 1998.

[3] C. Patrick Yue, Student Member, IEEE, and S. Simon Wong, Senior Member, IEEE, *On-chip spiral inductors with patterned ground shields for Si-Based RF IC's*, IEEE JOURNAL of solid-state Circuits, VOL.33, NO.5, MAY 1998.

[4] F. Rerat and J.F. Carpententier, *BiCMOS6M technology: high frequency characterisation of inductors and comparison with BiCMOS6 result*, EC98/34, Central R&D / Electrical Characterisation / Crolles / ST Microelectronics, 1998, Internal Report.

[5] Jianjun J. Zhou, Member IEEE, and David J. Allstot, Fellow, IEEE, *Monolithic transformers and their application in a differential CMOS RF Low-Noise amplifier*, IEEE JOURNAL of solid-state Circuits, VOL.33, NO.12, DECEMBER 1998.

[6] B. Murari, F. Bertotti and G.A. Vignole (Eds.), *Smart Power ICs Technologies and Applications*.

VLSI CMOS Design of a RD-CNN for Motion Control of Multi-Actuator Robotic Systems

Francesco Giuffrè

STMicroelectronics S.r.l.

Stradale Primosole 50, 95100 Catania - Italy

E-mail: francesco.giuffre@st.com

Introduction

The idea to electronically reproduce a number of particular cerebral capabilities, typical of living creatures, dates back some years. With regard to some logic-mathematical aspects, the elaboration speed of electronic machines is now vastly higher than any biological structure can sustain, but the ability to perform many other operations, even those which may appear simple and taken for granted, has not yet been reproduced by any machine or circuit, however complex it might be. Herein lies the interest of science and applied technology particularly in electronics and robotics, to seek to understand in depth the physical structure and basic biological mechanisms which occur in the control centre of every living creature: the brain.

In the field of systems and control engineering, considerable effort has been dedicated in the last years to the construction of robots endowed with an increasing degree of decision making and locomotive autonomy, capable of performing otherwise dangerous tasks for man. In particular, an interesting aspect is locomotion in robotic systems having a high degree of freedom and therefore numerous motor actuators.

Traditionally this problem has been resolved by resorting to complex control schemes, by a single processor which, in a sequential manner, organises the movement of each single actuator in order to achieve a harmonic and adjustable locomotion. This, however, does not occur in nature: also the least sophisticated creatures, from annelida or even amoeba manage to resolve the problem of locomotion brilliantly, in real time and in a highly adaptive manner, although they are still without a central nervous system but with a series of ganglia (nerve centres or neurons) distributed throughout the entire body.

In the D.E.E.S. laboratories of the University of Catania, interesting experiments have been carried out for the creation of biologically inspired locomotion schemes, with reference to the Cellular Neural Networks paradigm. This work is part of the above research program of which it represents one of the most advanced stages and regards the planning and "on chip" integration of a Reaction-Diffusion Cellular Neural Network (RD-CNN) for Motion Control, using STMicroelectronics 0.35 μm HCMOS6 technology, capable of working as "Central Pattern Generator" (CPG) [6], [9] for the locomotion of multi-actuator robotic systems.

Cellular Neural Networks (CNN)

General characteristics

The CNN is a n-dimensional array of mainly identical dynamic system, called cells, which satisfies two properties:

- most interactions are local within a finite radius r;
- all state variables are continuous valued signals.

Each cell is a dynamic non-linear, continuous time system, with a vector of inputs, one of states and another of outputs. In the traditional ANN, each cell receives in input in addition to its own, the outputs of all the cells forming the same network. CNNs are instead distinguished by the local connection, or rather the characteristic that the cell does not exchange signals directly with all the other cells in the network, but only with those in its 'neighbourhood', that is in a circumference (for planar networks) or in a spherical volume (for three dimensional networks) comprising only the units with which there is a connection. In spite of the "local nature" of the nearest neighbour interconnections, cellular networks are nevertheless imbued with some global properties because of the propagation effects of the local interactions during the transient regime.

The Nr neighbourhood, with respect to the array r, is defined as [2]:

$Nr(i,j,z)= \{C(h,k,l) \mid \max \{|h-i|,|k-j|,|l-z|\} \leq r\}$.

With:

$1 \leq h \leq m$; $1 \leq k \leq n$; $1 \leq l \leq p$; m, n, p = network dimensions.

The Figures shown below refer to a few common examples of planar and three-dimensional CNN [2]-[3]:

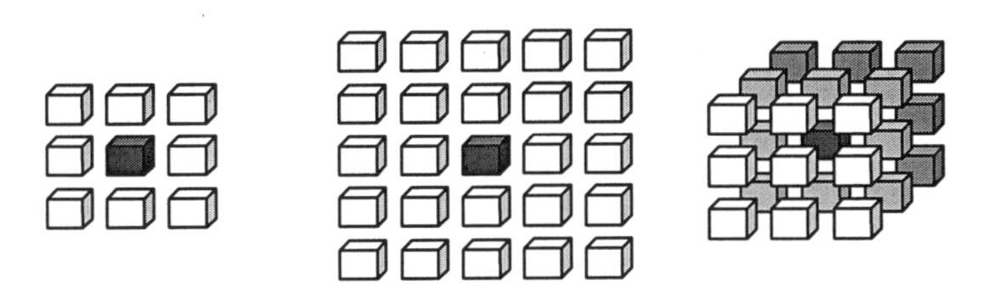

Figure 1: CNN Planar Neighbourhood r =1

Figure 2: CNN Planar Neighbourhood r =2

Figure 3: CNN 3-D Neighbourhood r =1

All inner cells of a CNN have the same circuit structure and element values. The inner cell is the cell which has $(2r+1)^2$ neighbour cells. All other cells are called boundary cells.

Each cell is made of a linear capacitor, a non-linear voltage-controlled current source, and a few resistive linear circuit elements. CNNs share the best features of the ANNs and Cellular Automata; its continuous time feature allows real-time signal processing found wanting in the digital domain and its local interconnections feature make it tailor made for hardware implementation.

One of the most appealing aspects of these circuits is just that they can be more easily implemented as VLSI devices than networks which are fully connected or whose coupling coefficients must be changed during a learning process. The combination of local connectivity, fixed templates (coupling coefficients), and feedback with simple, well defined dynamics at each node enables the circuits to be fabricated using standard micro-fabrication technology.

Although ANNs have an indisputably greater elaboration power, above all in particular 'learning' processes, thanks to their 'full connection', it is precisely this feature that renders them difficult to make [2]. On one hand there is a network typology (ANN) for which there are very powerful resolution algorithms, but which is difficult to implement in hardware, and on the other hand there is the CNN paradigm, which minimises the construction complexity by having a 'local connection', and seeks to resolve the same algorithms by adapting the hardware.

The structure of CNNs is particularly suitable for image processing task. Let us consider a two-dimensional array of data, such as a grey-scale digital image and let these data be the initial values for the cell state variables. As the CNN dynamic system evolves the state is changed according to the cloning templates. When the steady state is reached, the state or the output values represent the processed array of data, that is the processed image. CNNs are then used for noise removal, edge and corner detection, half-toning, motion sensitive filtering.

The characteristic of the CNNs regarding local connectivity is also the ideal key element for 'run-time' simulation of physical systems describable with partial differential equations (PDEs).

A spatial discretization is introduced by a grid. The values of the variables at the various positions are associated to the state variables of the corresponding cells into the grid. The spatial derivatives are approximated by finite differences while the time derivatives are left unchanged. In this way, it is often feasible to map these discretized PDEs into the CNN equations. Therefore the PDE is easily solved by the corresponding CNN evolution.

Many non-linear phenomena like solitons, autowaves, spiral waves, patterns etc. have been observed and studied by biologists, neurologists and physicians in non-linear active media or living structures, where transport processes take place, such as living neural tissues, physiological systems, ecosystems, as well as in chemical reactions or in combustion. They are often described by *Reaction-Diffusion Equations* [4] and the same phenomena have now been reproduced and observed in some CNN architectures. In this framework, CNNs represent then a powerful tool for their real-time simulation.

The R-D Equation is this type:

$$\frac{\partial U}{\partial t} = f(U) + \nabla^2 U$$

Where:

f(U)	is the reactive component;
$\nabla^2 U$	is the diffusive component

("∇^2" is the laplacian operator; "U" is a generic vector state variable).

The generation of "Turing patterns" [5]-[6]-[9] is related to this latter case, namely designs, reporting patterns or geometric forms, which are repeated in a more or less systematic manner. These patterns are in fact the result of a complex reactive-diffusive process with at least two different chemical species (activator and inhibitor) playing a part, which after a more or less long transient, reach a final state of equilibrium where there is a minimum of an energy function, depending in turn on many factors among which the boundary conditions, the form and dimensions of the space.

Typical representations of "Turing patterns" may easily be found in nature: the markings on the coats of some felines, a number of crystal formations, and in general where there has been the growth of biological or crystalline strata through the interaction and then diffusion of at least two different chemical species.

In this field, also the concept of "Autowave" [5]-[6] assumes great importance. They represent a particular class of non linear waves, which propagate without forcing functions, in strongly non linear active mediums. Their propagation takes place at the expense of the energy stored into the active medium; such energy is used to trigger the process into adjacent regions. In many cases the medium in which wave propagation occurs does not return to the original state (combustion), in others (nerve fibres, cardiac muscle, neurons), this occurs after a generally lengthy lapse of time. In all cases the passage between the two levels (high and low) of the active medium takes place with very different frequencies between them (slow-fast systems).This particular wave is endowed with the following features:

- propagation in non-linear active medium, sustained by energy released from the same medium;
- absence of reflection;
- constant amplitude and form in time;
- annihilation of the wave fronts in the case of mutual collision.

The realisation by CNN in this case allows to have cells which alter their condition between two extreme values in continuous mode and without alteration in amplitude.

Before proceeding further, it seems opportune to analyse in greater detail the basic unit (cell), often referred to above, which makes up a CNN.

Analysis of a Neural Cell

In its simplest form, a cellular, single layer neural cell, may be represented by the Scheme shown in the Figure 4.

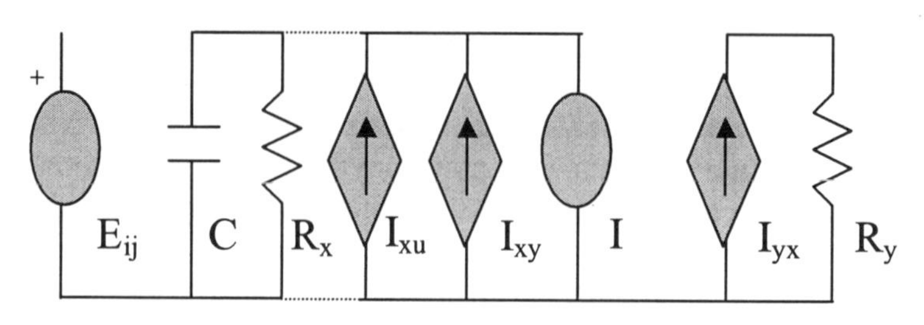

Figure 4: Single layer CNN cell

The equations of this circuit are:

$$y_{ij} = f(x_{ij}) = 0.5 \cdot (|x_{ij} + 1| - |x_{ij} - 1|)$$

$$C \cdot \dot{x}_{ij} = -\frac{1}{R_x} x_{ij}(t) + \sum_{C(k,l)\in N_r(i,j)} \hat{A}(i,j;k,l) y_{kl} + \sum_{C(k,l)\in N_r(i,j)} \hat{B}(i,j;k,l) u_{kl} + \\ + \sum_{C(k,l)\in N_r(i,j)} \hat{C}(i,j;k,l) x_{kl} + I$$

$$N_r(i,j) = \{C(k,l) | \ \max(|k-i|,|l-j|) \le r\}$$

the suffices u, x, and y denote the input, state and output respectively. The node voltage V_{xij}, of C(i, j) is called the state of the cell.

The node voltage, called Vuij, is the input of C(i, j). The node voltage V_{yij} is called the output.

We can generalise the single-layer CNN to a multi-layer CNN.

Instead of only one state variable in the single-layer case a multi-layer CNN (or MCNN) will be composed of cells having several state variables, one for each layer. The interaction between the state variables of the same cell can be complete while the cell-to-cell interaction remains local.

Our aim was to implement a CNN able to represent RD equations. In particular both the reactive part and diffusive one were designed in [6]. The dynamic of the single cell of the two layers CNN used, is reported in the block scheme of Figure 5.

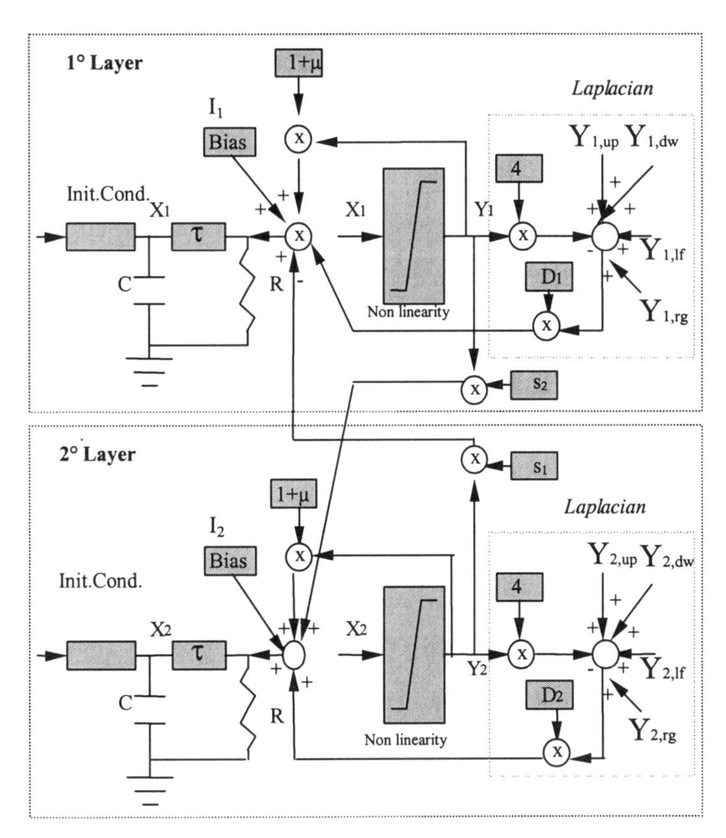

Figure 5: Two layer CNN cell

In the dotted boxes, the dynamics of each layer of the single cell is represented. In particular:

- μ, ε, D_1, D_2, s_1, s_2, I_1, I_2 are variable parameters to be fixed;
- X_1 and X_2 are the state variables of the 1th and 2nd layer, of the cell located at (i, j);
- Y_1 and Y_2 are the outputs of the cell located at (i, j).

It is evident that also cells in a network which are not included in the vicinity of a given cell, are still equally influenced by it, not in a direct way as occurs for all those within a boundary but in a transitive fashion through other cells, because of the propagation effect of the continuos-time dynamics.

While the two layers interact within each single cell generating on the oscillatory, slow-fast behaviour described in [6], conversely, the interaction with the neighbouring cells is obtained separately by means of two circulant diffusion templates with diffusion coefficients D_1 (for the first layer) and D_2 (for the second one) respectively, which determine the entity of reciprocal influence between adjacent cells.

The I_1 and I_2 parameters, known as "Bias" [3] on the other hand determine the point in which the weighed sum at the input to the cell varies the polarity of the state X and may be considered to all intents and purposes as further inputs. The bias current plays a crucial role in determining the behaviour of the CNN array.

In the following, the so-called Zero-Flux (Neumann) boundary condition will be assumed.

The important thing to observe is that, by appropriately fixing μ, ε, D_1, D_2, s_1, s_2, I_1, I_2, entirely different behaviours may be obtained among cells and in particular also those which lead to the generation of "Turing Patterns" and "Autowaves" [7]-[8].

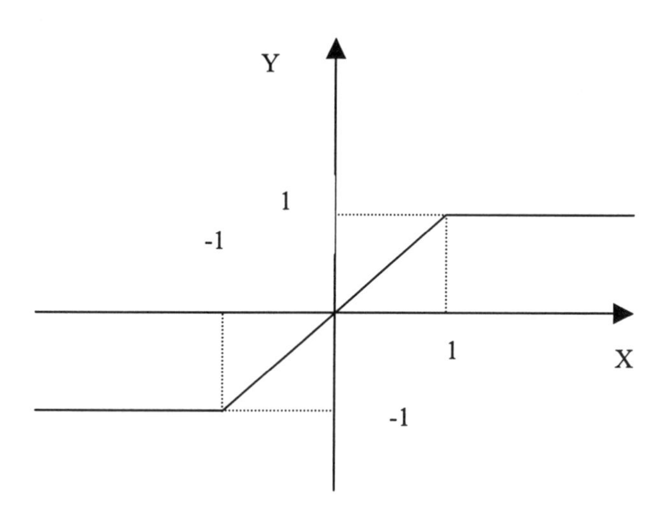

Figure 6: Output of a single cell with double-level saturation region

The state and output of each single cell have an analog dynamic. In particular the output of $Y_{i,j}$ is not linear and closely resembles a sigmoid [2]-[3], with one transition ($|X|<1$), known as an α type, and two saturation areas ($|X|\geq 1$) known as β type. The cell has at most 2 stable equilibrium points.

These are points where the circuit state will settle after the transient has decayed. We use a slope of 1 in the transition region.

The CNN can be described by matrices (cloning templates) and used as a programmable processing device where the cloning templates represent the elementary program of the CNN (instructions or subroutines). This is the idea behind the so-called CNN Universal Machine (CNNUM) [10], where the adjective universal must be intended in the sense of Turing: it is able to realise any conceivable algorithm.

It is now clear that the CNN behaviour is basically dictated by the templates. However the choice of the templates suitable to obtain a desired processing task is hard to be accomplished.

To complete the concepts outlined in Figure 6 it should be remembered that cells capable of giving multi-level are being studied [3]; the simplest method to obtain this result is to allow the piece-wise-linear curve to have more than two segments of zero slope, adding additional steps to the sigmoid.

In this case the previous Figure would be modified as the following illustrates:

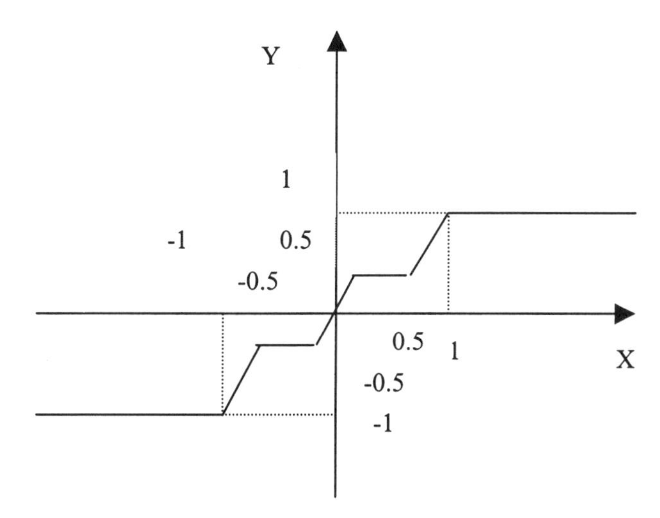

Figure 7: Output of a single cell with multi-level saturation region

The number of the equilibrium states (horizontal lines) may theoretically have any value, but in reality this is drastically limited by the tolerance of the components used and by external interference and noise.

Analysing Figure 5, let us consider the sub blocks inside each block, called "Laplacian". They represent the diffusive part connecting each cell with the neighbours. In particular this diagram refers to the case in which all the neighbours cells are present. In cases where the neighbouring cells are fewer, the coefficient must be accordingly reduced (3, 2, 1, 0 for a number of cells varying for 3 to zero respectively).

Project for the Integration of an RD-CNN for Motion Control of Multi-Actuator Robotic Systems

Aims and details of project

The aim of this work is to plan and integrate an RD-CNN composed of a 16 x16 matrix of double layer cells on silica for motion control of multi-actuator robotic systems. The design is also able to accomplish more general PDE's solution.

A 0.35 μm technology, called hcmos6, by STMicroelectronics, was chosen. The project requirements may be summarised as follows:

- double layer CNN, in order to be able to generate complex phenomena such as Autowaves and Turing patterns;
- programmable parameters in suitable ranges able to generate complex dynamics as specified above;
- programmable connections to enable different spatial geometry networks;
- programmable run time of oscillations in each cell.

Design Stages

When designing a VLSI implementation of any circuit, it is helpful to be able to scale the component values to accommodate the fabrication technology, power dissipation, and parasitics.

In particular for this project, the first step was to adjust or scale down the original equations of a single cell, to make them suitable for the hcmo6 technology employed, starting from the desired characteristics of the non-linear output function. Since the time constant of the cell is dominated by R·C, it is important to be able to adjust this so that it is slow compared to parasitic time constants (for instance, to make C significantly larger than parasitic capacitances). The aim of the second stage was to break a single cell down into its fundamental parts, followed by the planning and the first series of simulations of the above blocks, considered initially as separate from each other.

The third step was to connect the various blocks together and carry out the first simulations of the complete cell to check its correct working.

Having obtained the desired behaviour of a single cell, the project proceeded with real and proper simulation of the RD-CNN, considered at this time as formed by a 6x1 array of cells. The results of these simulations, which proved decisively positive, will be shown, thus demonstrating the actual of the desired reaction-diffusion phenomenon.

Scaling the equations

A fundamental and top priority operation was to adapt the equations referring to standard RD-CNN to our technology. The following procedure was used beginning with the equations; here reported for the art of completeness:

$$y = \frac{1}{2}\left[|x+1| - |x-1|\right]$$

$$\begin{cases} \dot{x}_1 = -x_1 + (1+\mu)y_1 - sy_2 + I_1 + D_1[\nabla_*^2\, y_1] \\ \dot{x}_2 = -x_2 + (1+\mu)y_2 + sy_1 + I_2 + D_2[\nabla_*^2\, y_2] \end{cases}$$

where the term $\nabla_*^2\,(\,)$ denotes the discretized laplacian operator [6].

By comparing these equations with those of a cell in which the nonlinearity is represented by the following Figure:

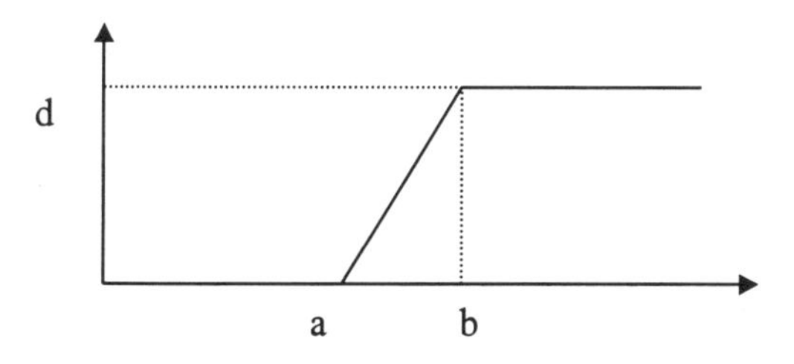

Figure 8: Cell output

we obtain

$$\begin{cases} C\dot{x}_1'' = -R^{-1}x_1'' + (1+\mu)y_1''' - sy_2''' + I_1^B + D_1[\nabla_*^2\, y_1] \\ C\dot{x}_2'' = -R^{-1}x_2'' + (1+\mu)y_2''' + sy_1''' + I_2^B + D_2[\nabla_*^2\, y_2] \end{cases}$$

where

$$R = \frac{2G}{d}\ ;\ I_i^B = I_i' R^{-1};\ G = \frac{b-a}{2}\ ;\ y' = \frac{1}{2}\left[\left|x'' - h + G\right| - \left|x'' - h - G\right|\right]$$

$$y_i'' = y_i' + G\ ;\ y_i''' = y_i''\frac{d}{2G}$$

$$I_1' = h - (1+\mu)G + sG + GI_1 = h + (I_1 + s - 1 - \mu)G$$

$$I_2' = h - (1+\mu)G - sG + GI_2 = h + (I_2 - s - 1 - \mu)G$$

and the relations connecting $<x''>$ to $<x>$ and $<h>$ to $<b>$ and $<a>$ are:

$$h = G + a = \frac{b+a}{2}; \qquad x'' = Gx + h$$

Given that for technical reasons has been necessary to restrict the variation range of the variable states between 0.7 e 3 the following limitations (see Figure 9) of possible parameter values may be calculated:

$$\begin{cases} b > a \\ b < 2a - 0.7 \\ a < 1.602 \end{cases} \text{ or } \begin{cases} b > a \\ b < \dfrac{a+6}{3} \\ a > 1.602 \end{cases}$$

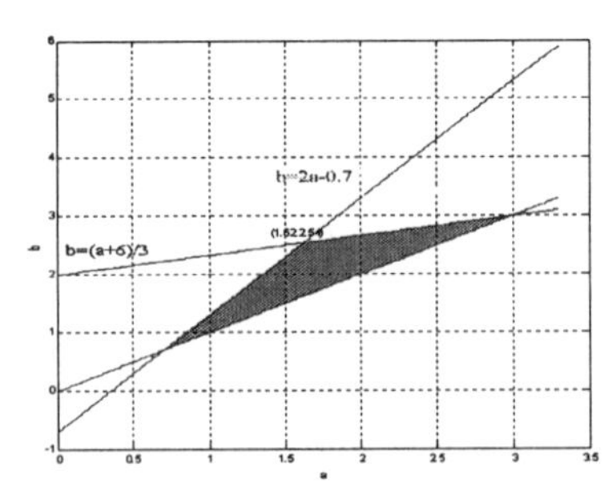

Figure 9: Acceptable range of variables

An initial choice of parameters, encouraged by the simulations performed, furnished the following values: a = 2 V; b = 2.44 V; G = 220mV; h = 2.22 V; R = 44kΩ; I_1 = 45.4μA; I_2 = 38.5μA;

Subdivision into fundamental logic blocks

The fundamental logic blocks making up a single cell are:

- initial condition block;
- block defining equation coefficients;
- R-C group;
- delay block;
- bias current generators;
- multiplier block;
- adder block;
- non-linear output block.

The circuit design will be shown for each of these blocks and an explanation of the working will be given, bearing in mind that they are surely susceptible to changes aimed at optimising performance and reducing the overall area of silicon.

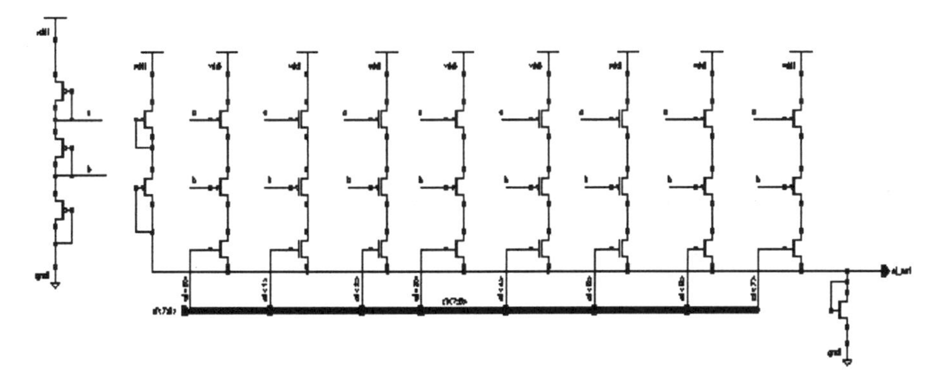

Figure 10: Initial condition block

Initial condition block

The initial conditions, or rather the voltage at time equal to zero on the capacitors, are fundamental for a correct evolution of the reaction-diffusion phenomenon. Each voltage is fixed with an array of 9 half-mirror current generators with different multiplier factors, one of which with a preestablished current so that a minimum voltage in the chosen range is provided. The other 8 are enabled with a bit and sized in such a way that if they are all activated, they give the maximum value of voltage in the pre-selected range; every enabling byte provides all the intermediate values between the selected limits.

Coefficients block

The vector coefficients μ, ε, s, I, D, which in the general RD-CNN equation refer to voltage and current, are fixed by enabling bits already seen in the previous block and which will also be found in successive blocks. The bits, grouped in enabled bytes, are fixed in this block.

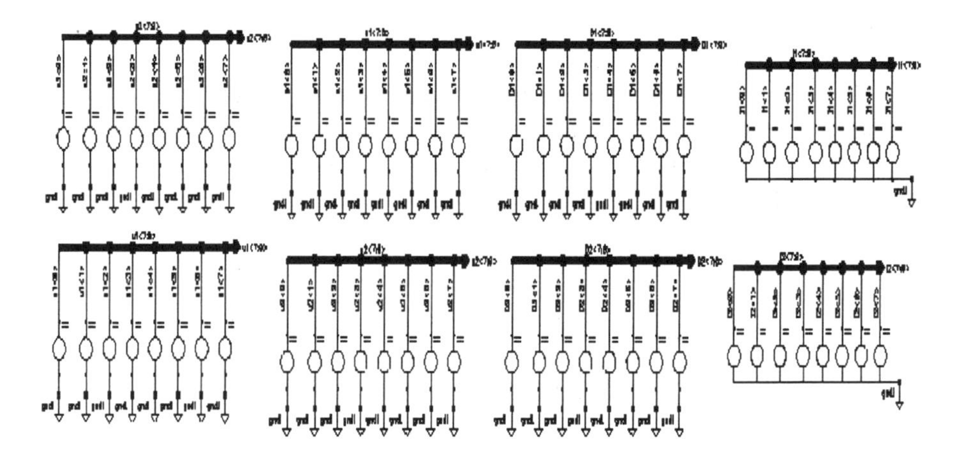

Figure 11: Coefficient blocks

R-C Group

The value of the only resistance present was fixed at 44kΩ after scaling the equations. It was originally hoped to integrate the capacitor as large as possible but its value always fluctuated around some tens of pF. The time constant of the circuit constituted by the R-C block defined above, proved very low and this indicated an oscillation frequency of some hundreds of kHz, entirely unsuitable for the application for which it was designed.

Delay block

Attempts were made to solve the problem of excessive oscillation frequency by introducing a delay block, comprising a transconductance amplifier according to the following Diagram:

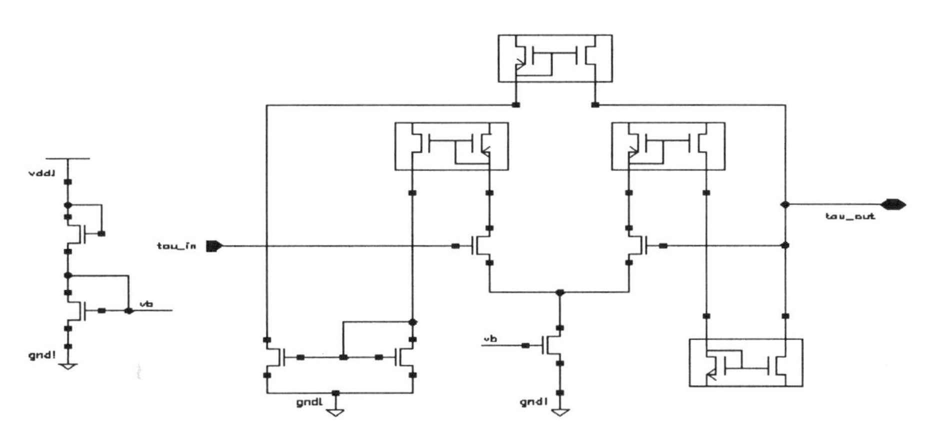

Figure 12: Delay block

whose sub-blocks are of the P and N current mirrors:

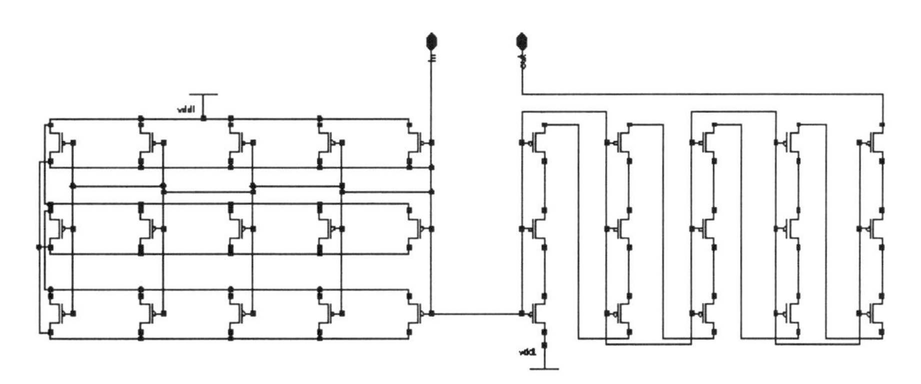

Figure 13: 225/1 P current mirror

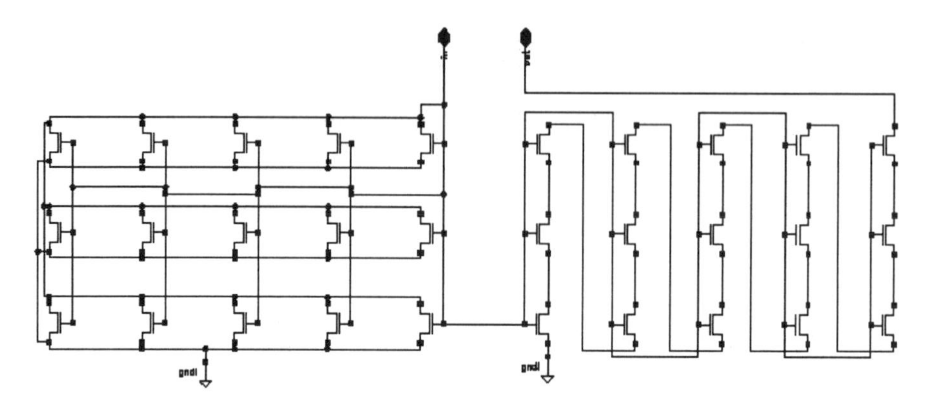

Figure 14: 225/1 N current mirror

Bias current generators

The I parameter appearing in the RD-CNN equation which represents the Bias current is fixed and, when necessary, made to vary within a pre-fixed range by the scheme indicated below.

The function performed by the enabling bits is the same as that previously described for the initial condition block.

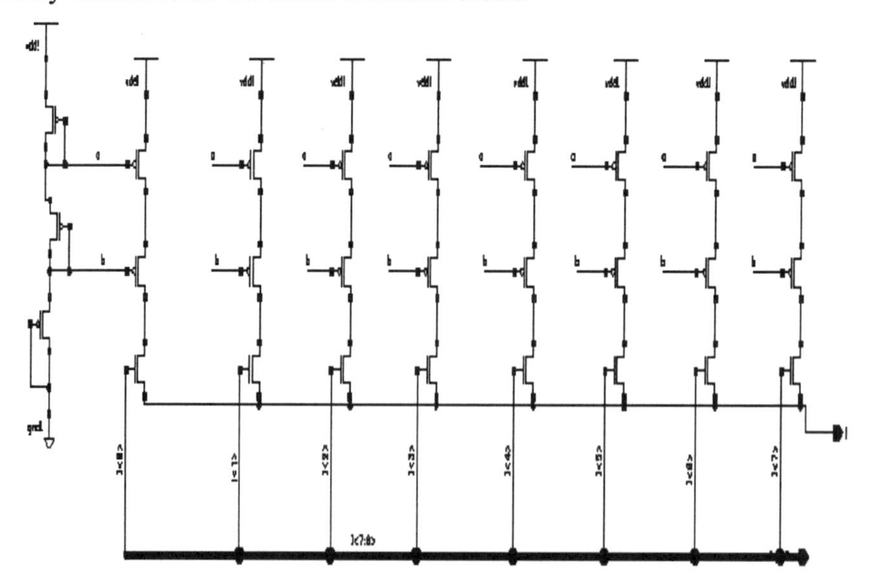

Figure 15: Bias current generator block

Multiplier block

The following type of multiplication are present in the RD-CNN equation: $(1+\mu+\varepsilon)*Y$, $s*Y$ e $d*\Sigma Y$. Due to the diverse range of possible values which the coefficients '$(1+\mu+\varepsilon)$' and 's' on one hand and 'D' on the other may assume, and since it is desirable to have the possibility to vary such coefficients in a determinate range (a programming requirement), given that these ranges prove to be incompatible with each other, it was thought to implement two different kinds of multiplier whose layout is shown here:

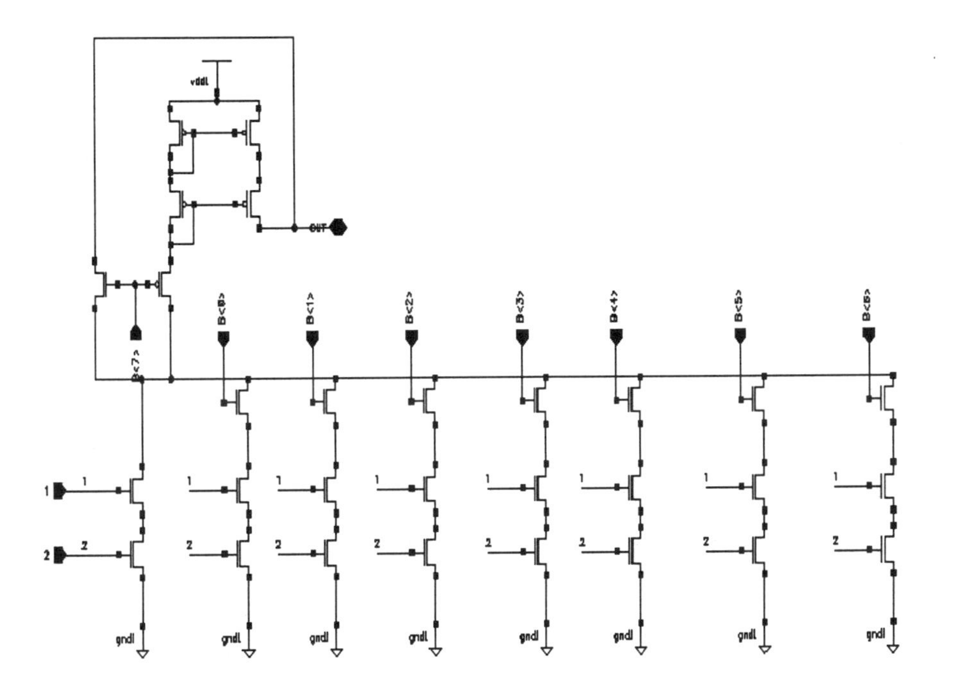

Figure 16: Multiplier for '$(1+\mu+\varepsilon)$' and 's' coefficients

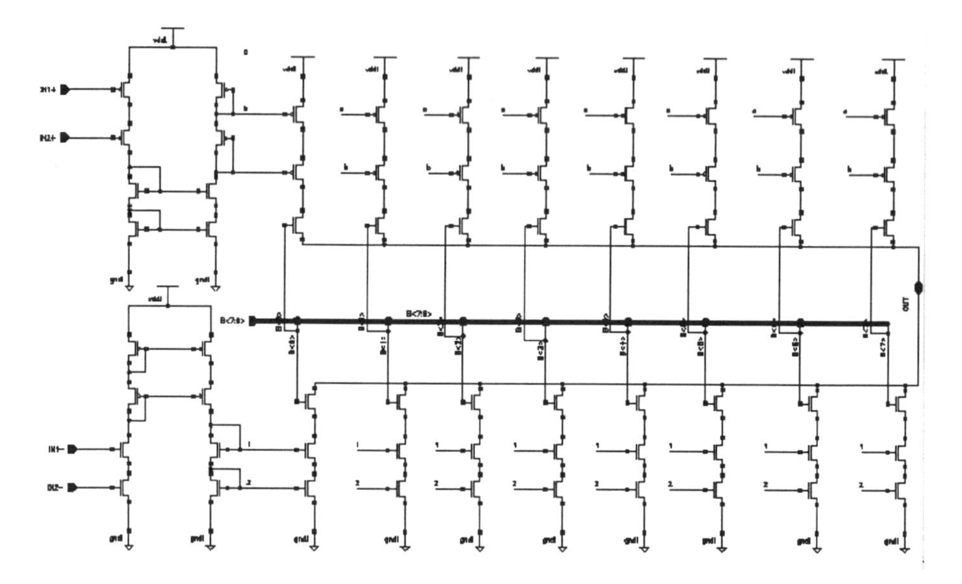

Figure 17: Multiplier for 'D' coefficients

Adder block

This block performs the algebraic adding operation of the output of the neural cells adjacent to the considered one and of the output of the same cell multiplied by 4, since that is the number of neighbouring cells. The result of this operation is then multiplied by the diffusion coefficients. The layout is shown in the Figure 18.

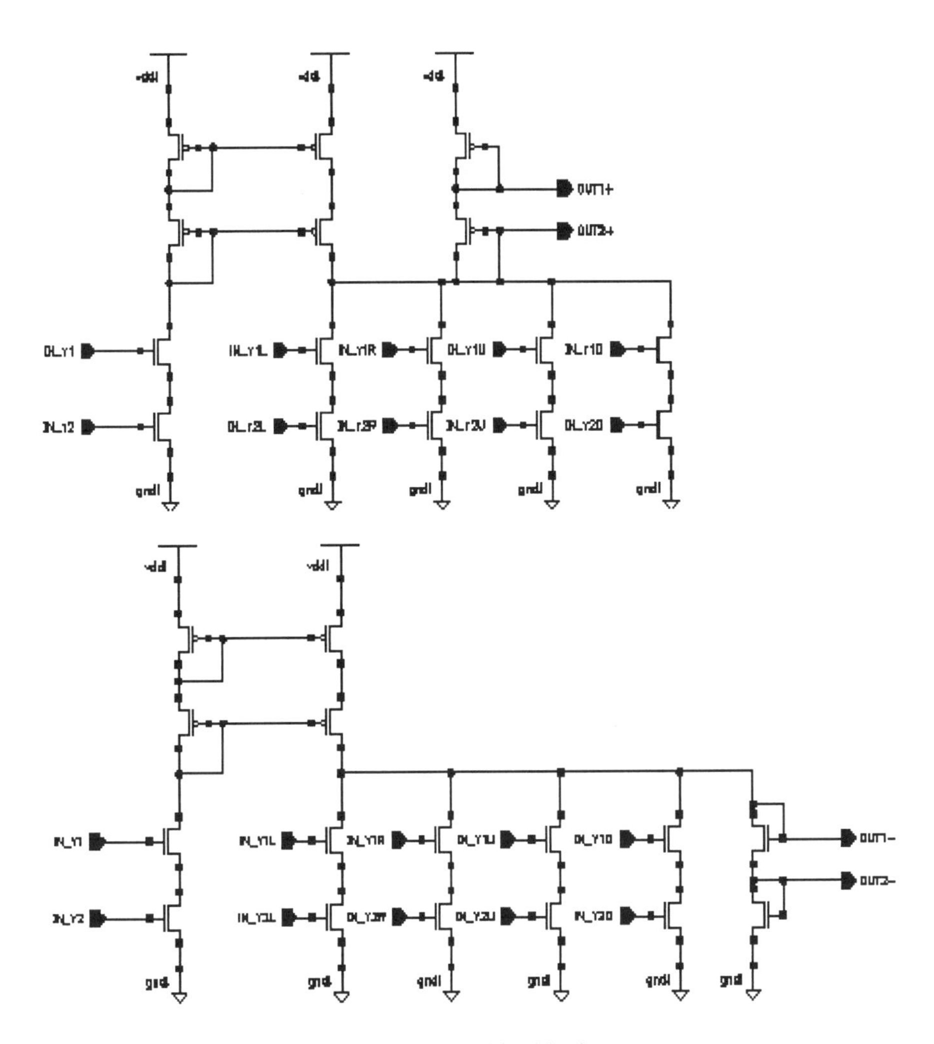

Figure 18: Adder block

Non-linear output block

One of the most important and certainly most critical parts appeared immediately to be the output block, formed by the part which generates, beginning from the state of the cell, the output non-linearity. The slope of the line, within the α region and the value of the output saturation current of the β region, were fixed at values deriving from the considerations outlined previously in the equation adjustments, keeping in mind not to excessively waste the area. The diagram of this block is reported below:

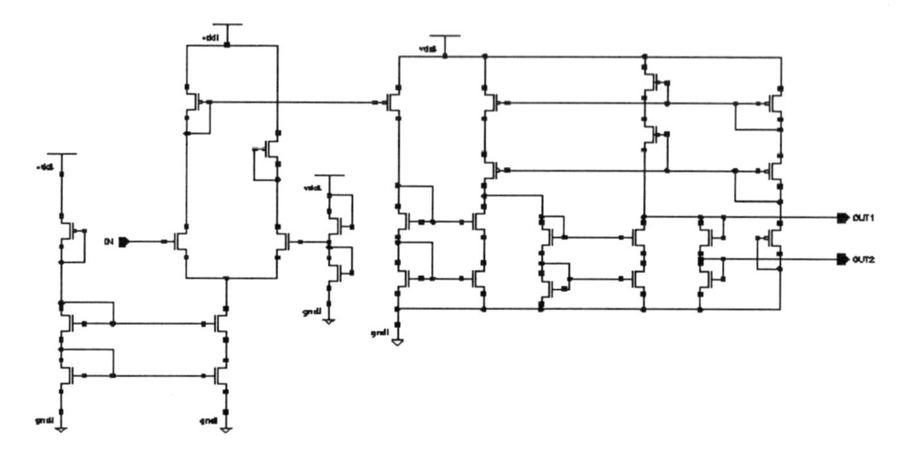

Figure 19: Non-linear block

Complete Scheme of a neural cell

Having defined its various parts, it is possible to have a complete vision of the neural cell.

Among the initial condition blocks, the R-C group and the delay block, MOS in 'transmission gate' formation are present which have the task of maintaining, driven by a clock, the initial conditions for a specific time before allowing the evolution of the cell dynamics. The complete Scheme is as follows:

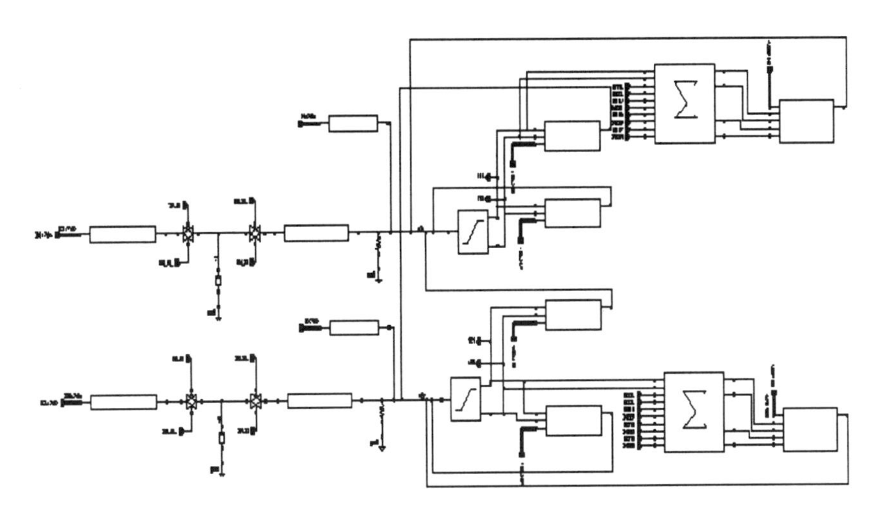

Figure 20: Complete scheme of a single cell

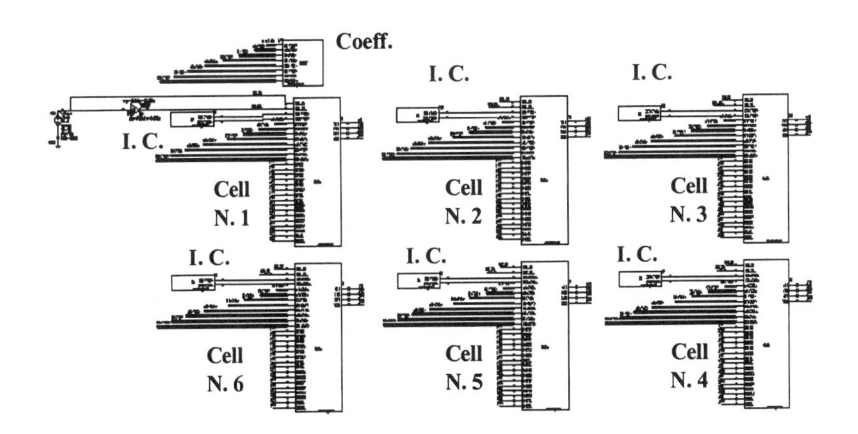

Figure 21: Array 1x 6 of Neural Cells

Layout of an RD_CNN composed of 6x1 cells

By bringing the initial conditions block outside of each single cell, given that it differs from one cell to another one, considering the cell in turn as a unique block and by connecting them appropriately, the final architecture shown in the Figure 21 is obtained.

Final simulation

Simulation of the entire RD-CNN shows the desired features, namely a slow-fast type phenomenon, with a compatible oscillation period with the specifications for which it was designed.

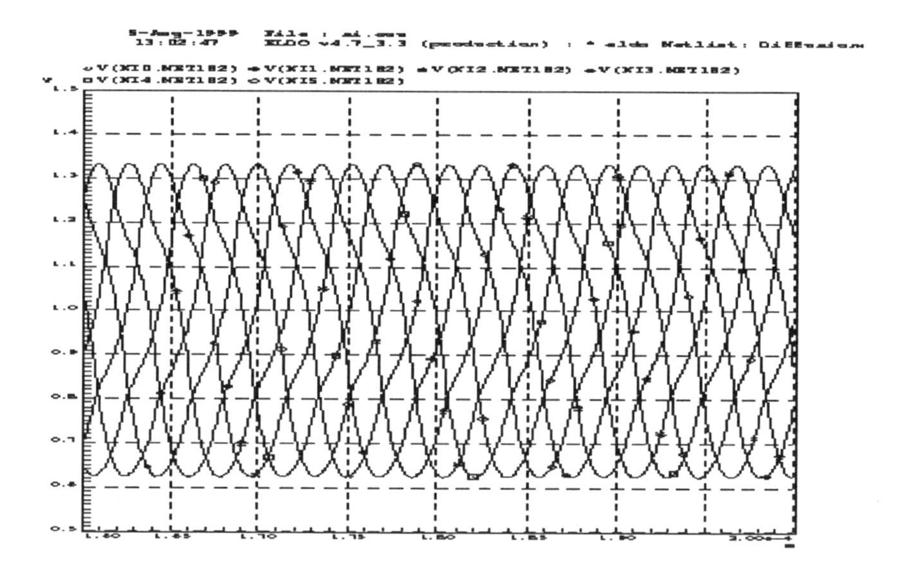

Figure 22: Output of a RD-CNN (array 1 x 6)

Conclusions

The design of a double layer CNN, for Motion Control of multi-actuator robotic systems and, in general way, for PDE solution has been shown. Definition and implementation of almost all the functional blocks in HCMOS6 technology, of which a single cell is composed, has been followed by a behavioural verification phase by simulating with many positives results.

The project then will proceed with the resolution of a number of problems encountered and the optimisation of the solutions found. It is possible, for example, to opt for a halving of the enabling bits which define the coefficient values, in such a way to have a notable saving of space, to the detriment of a slight loss of precision with regard to the theoretically desired values. The question of what is reasonable to render programmable within the network and also within each single cell, remains open: many coefficients, for instance, may be made constant, provided that the specification envisages the possibility to generate the complex dynamics in consideration (Autowaves and Turing patterns). The problem concerning the enormous number of pins that the final integrated circuit should theoretically have must still be defined and resolved, and for this reason it is thought that it is necessary to resort to signal multiplexing. The type of memory to be used is still to be finalised, as is the actual programmability of the connections which should allow maximum elasticity in various network layouts. The actual software simulation of the delay blocks, requires further testing aimed at real implementation.

In its final version, the work presented here will in all likelihood represent a reference point for the ever more numerous work having the objectives of studying, modelling and reproducing the complex phenomena present in nature, especially for the biological neural systems. Nature has spent millions of years refining and putting the finishing touches to her work, achieving a very high level of perfection. The challenge to model cerebral structures has just begun, but common efforts and the progress to date in all branches of research are contributing to opening up complicated animal neural-biological mechanisms, thereby offering the opportunity to embark on the road which will lead to obtain such sophisticated and powerful electronic structures as to render the behaviour of a machine ever more similar to that of living beings.

References

[1] J.J. Hopfield, *"Neural networks and physical system with emergent computational abilities"*, Proc. Natl. Acad. Sci. USA, vol. 79,1982.

[2] L.O. Chua and L. Yang, *"Cellular Neural Networks: Theory and Applications"*, IEEE Trans. On Circuits and System, vol. 35, n. 10, 1988.

[3] L.O. Chua, L. Yang, K.R. Krieg, *"Signal Processing Using Cellular Neural Networks", Dept. Of EECS, University of California, Berkley, 1991.*
[4] V.I. Krinsky, *"Autowaves: Result, Problems, Outlooks in Self-Organization: Autowaves and Structures Far from Equilibrium"*, Springer-Verlag, Berlin, 1984.
[5] P. Arena, S. Baglio, L. Fortuna, G. Manganaro, *"Complexity in a two layer CNN"*, proc. CNNA '96, Seville-Spain, 1996.
[6] P. Arena, M. Branciforte, L. Fortuna, *"A CNN experimental Frame for Patterns and Autowaves"*, Int. Journal of Circuits, Theory and Applications, 1999, in press.
[7] P. Arena, S. Baglio, L. Fortuna, G. Manganaro, *"Self organization in a two layer CNN"*, IEEE Trans. On Circuits Systems, part I, Feb. 98.
[8] P. Arena, L. Fortuna, G. Manganaro, *"A CNN for Pattern formation and active wave propagation"*, European Conf. on Circuit Theory and Design, ECCTD '97,Budapest, 1997.
[9] A. M. Turing, *"The chemical basis of morphogenesis"*, Phil. Trans. R. Soc. London, vol. B327, pp. 37- 72, 1952.
[10] T. Roska, L.O. Chua, *"The CNN Universal Machine: An Analogic Array Computer", IEEE* Trans. on Circuits and Systems, Analog and Digital signal processing, vol. 40 no.3, March 1993.

CNN Based Integration Design in 0.35µm CMOS Technology for Data Protection

Luciano Salvatore Petralia

DEES, Engineering Faculty, University of Catania

Viale A. Doria 6, 95125 Catania - Italy

Introduction

CNN (that stands for *Cellular Neural Network,* or *Cellular Non-linear Network* according to the recent definition provided from one of the inventors, Leon Chua) is a class of simple, non-linear coupled dynamic analog circuits used to process a large amount of information in real time.

They were introduced in 1988 by the above mentioned Chua and L. Yang, with their papers "Cellular Neural Networks: Theory" and "Cellular Neural Networks: Applications" on the IEEE transactions on Circuits and Systems journal.

The original model was subsequently modified by addition of other analog or digital components, but it remains the most widely used one due to fact that it offers to a good compromise between simplicity and versatility, and is also easy to implement.

The fundamental building block of the CNN is the cell, a lumped circuit containing both linear and non-linear elements, whose regular repetition forms an array of cells.

Each cell is coupled only to its neighbouring cells. Adjacent cells can interact directly, while more distant cells can have an indirect influence due to propagation.

Due to the local connectivity and to the fact that all the cells are equal, we can think to implement them in a VLSI technology.

A first realisation on chip in a CMOS technology was made in 1992 by Chua and Cruz. The circuit has very interesting properties: we can extrapolate[1] a performance of 0.3 TeraXPS[2] on a chip 10x10 mm^2 with a technology using MOS with a channel length of 2 μm.

Chua e Roska presented an article about the CNN Universal Machine (CNNUM), a circuit with an integration density of 33 cells for mm^2.

The CNN applications widen from image processing, data protection (cryptography), resolution of partial differential equations (such as the diffusion equation, the Laplace and the wave equations and so on) and, with a modified structure, to Neurobiology, with the motion control and the patterns recognition.

The Chua and Yang CNN architectural model

The CNN model proposed by Chua e Yang consists of an array of cells (also referred to as neurons or nodes), each of them containing both linear and non-linear elements, all cells are supposed equal and the array is a rectangular one with the cells distributed on a regular grid.

Moreover each of them is locally connected to all the neighbouring cells it, but an indirect influence with no adjacent elements is assured thanks to dynamic propagation.

Let us consider an MxN two-dimensional array CNN and let us call C(i,j) the generic cell placed on the ith row and the jth column. In Figure1, we consider a 4x4 CNN array:

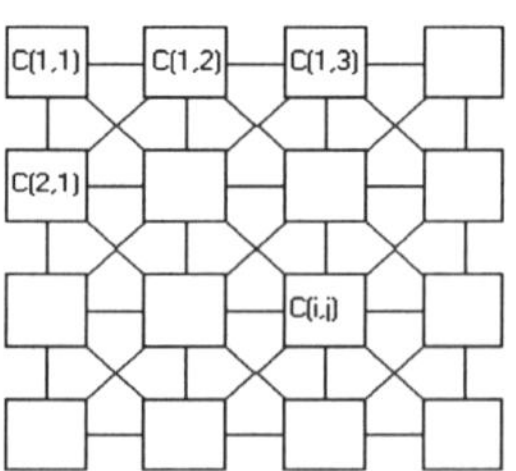

Figure 1: A 4x4 CNN

[1] Cause the analogic elaboration, it's no exact to refer to the instruction per second completely. The value fornished means the same instruction processed by a digital calculator has the same performance index.

[2] 0.3 TeraXPS = 0.3 x 10^{12} Instructions per second = 300000 MIPS.

Of course it's possible to define arrays of any dimension but for the sake of simplicity we restrict ourselves to two-dimensional arrays.

Now we need some definitions about CNN, whose validity is general for all the proposed models from the earlier to the latter ones.

Definition 1

We define r-neighbourhood set of C(i,j) the set:

$$N_r(i,j) = \{C(k,l) | \max(|k-i|, |l-j|) \leq r \text{ , } 1 \leq k \leq M \text{ ; } 1 \leq l \leq N\} \quad (1)$$

where r is the so called neighbourhood radius. A 3x3 array is referred to as an array with neighbourhood radius one, a 5x5 array as an array with neighbourhood radius two and so on. We note that the following property is valid for the cells: if C(i,j) $\in N_r(k,l)$ then C(k,l) $\in N_r(i,j)$.

Now we can observe the scheme of the basic cell. In Figure 2 we have the circuit equivalent, in which we may point out all the elements that describe CNN behaviour:

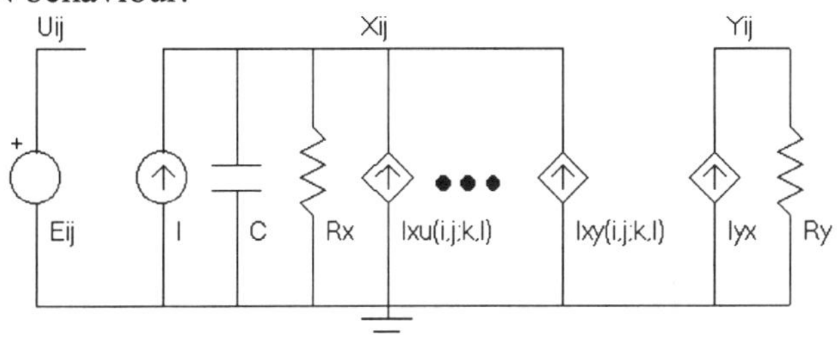

Figure 2: Circuit scheme of the basic cell C(i,j)

Here *u, x, y* denote respectively input, the state and the output of the neuron respectively. The voltage v_{xij} is the *state* of C(i,j) and we suppose that the absolute volume of the initial condition on it to be smaller or equal to one.

Voltage v_{uij} is the *input* of C(i,j) and we consider it as a constant whose module is smaller or equal to one. Finally the voltage v_{yij} is the *output* of C(i,j).

Moreover let us consider the elements of such a circuit. We may find E_{ij}, an independent voltage source, an independent current source I, a linear capacitor C, two linear resistances R_x ed R_y, and at least *3m* voltage controlled current sources (VCCS), which are the coupling terms with the neighbouring cells, by the controlling input voltages v_{ukl} and by the feedback of the controlling output voltages v_{ykl} of the *m* neighbouring cells.

Now we give the equations for these controlled current sources, in which we may note the dependence from the controlling voltages. In particular we have that for the two current sources $I_{xy}(i,j;k,l)$ e $I_{xu}(i,j;k,l)$ the following equation holds:

$$\begin{aligned} &I_{xy}(i,j;k,l) = A(i,j;k,l)v_{ykl} \\ &I_{xu}(i,j;k,l) = B(i,j;k,l)v_{ukl} \\ &\forall\ C(k,l) \in N_r(i,j) \end{aligned} \tag{2}$$

The only non-linear element of the cell is a piecewise linear current source I_{xy} whose equation is:

$$I_{yx} = \left(\frac{1}{R_y} \right) f(v_{xij}) \tag{3}$$

the following graph gives us the characteristic of $f(v_{xij})$:

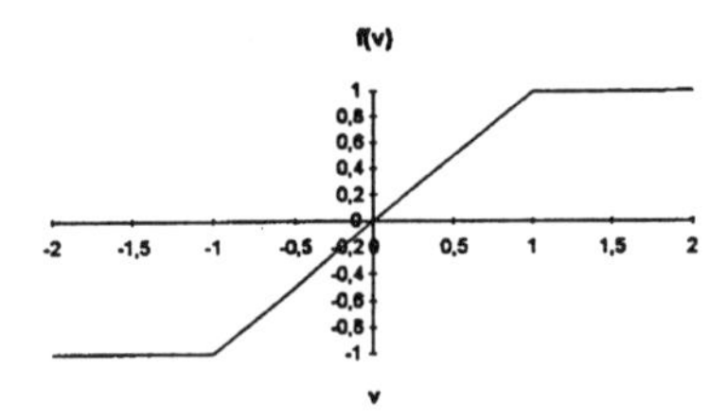

Figure 3: Characteristic of piecewise-linear current source

In equations (2) the coupling coefficients $A(i,j;k,l)$ and $B(i,j;k,l)$ appear. They are called the cloning templates, and more precisely the feedback template and the control template respectively.

Definition 2

Let us consider the non linear function f(v) of variable v. It's analytically expressed by the following equation:

$$f(v) = \frac{1}{2}\left[|v+1| - |v-1| \right] \tag{4}$$

From the observation of the circuit of Figure 2 the equations of the cell easily derive. In a more complete version the state coupling contribution can be taken into consideration, giving rises so called State Controlled CNN (SC-CNN) by the additional state template whose coefficients are the $C(i,j;k,l)$:

$$C\frac{dv_{xij}(t)}{dt}=-\frac{1}{R_x}v_{xij}(t)+\sum_{C(k,l)\in N_r(i,j)}\hat{A}(i,j;k,l)v_{ykl}(t)+ \tag{5.a}$$

$$+\sum_{C(k,l)\in N_r(i,j)}\hat{B}(i,j;k,l)v_{ukl}(t)+\sum_{C(k,l)\in N_r(i,j)}\hat{C}(i,j;k,l)v_{xkl}(t)+I$$

This equation is completed by a series of other equations, that define the output, the input behaviour and the boundary conditions, and they are precisely:

- The Output equation:

$$v_{yij}(t)=\frac{1}{2}\left[\left|v_{xij}(t)+1\right|-\left|v_{xij}(t)-1\right|\right] \tag{5.b}$$

- The Input equation:

$$v_{uij}=E_{ij} \tag{5.c}$$

- The Constraint equations:

$$\left|v_{xij}(0)\right|\le 1$$

$$\left|v_{uij}\right|\le 1 \tag{5.d}$$

$$1\le i\le M; 1\le j\le N$$

- Parameters assumptions:

$$\hat{A}(i,j;k,l)=\hat{A}(k,l;i,j),\quad 1\le i,k\le M, 1\le j,l\le N$$

$$C>0, R_x>0 \tag{5.e}$$

Remarks:

1. All the CNN *inner cells* have the same structure and the same template element values. An inner cell has $(2r+1)^2$ neighbouring cells, where r is the neighbourhood radius. Other cells are called the *boundary cells*.
2. *C, R_x, R_y* can be conveniently chosen by the designer, taking into consideration the problem of the power dissipation. *CR_x* determines the time constant of the system, even if, the circuit being non-linear, it is not rather appropriate to call it time constant, but we have values in the range of 10^{-8}, 10^{-5} s.
3. We said that the CNN dynamic depends upon the template element values. It is important to point out that there are some elements of the feedback template whose sign is positive: $A(i,j;k,l)>0$. This is important to ensure that the output settles to a well defined value, even when using non-ideal circuit elements.

4. The assumption of condition (5.e), that is $\hat{A}(i,j;k,l) = \hat{A}(k,l;i,j)$, is called as *reciprocity condition*. It is not always verified, because of circuit elements tolerances, or because intentionally we want to introduce certain characteristics. When the reciprocity condition hold we tell about *reciprocal templates*.
5. The I term in the (5) equations, derives from an independent constant current source. It is the so called **bias** or **threshold** of the cell, and its aim is to bring the state variables of the cell towards well defined equilibrium points or towards well defined trajectories of the state space.

We'll suppose always verified an hypothesis (except when specified), the hypothesis that the templates are *space-invariant*, that is all the templates have the same values for each of the cells in the array. This concept can be mathematically expressed by the following relation:

A(i,j;k,l)=A(k-i,l-j) and B(i,j;k,l)=B(k-i,l-j).

We have to do further considerations about CNNs, especially about the state variables and the input and output ones.

All the CNN states v_{xij} are bounded for all time $t > 0$ and the bound v_{max} can be computed by the following formula for any CNN [1]:

$$v_{\max} = 1 + R_x|I| + R_x \cdot \max_{1\le i\le M, 1\le j\le N}\left[\sum_{C(k,l)\in N_r(i,j)}\left(|A(i,j;k,l)| + |B(i,j;k,l)|\right)\right] \tag{6}$$

This is important for circuit design purpose, because we can understand the limits inside which the voltages change, and how we have to scale or to set the circuit parameters to satisfy the design specifics.

Anyway, we have to say that to verify this result we need to choose the inputs v_{uij} as finite constants, as we saw in the Input equation (5.c).

Moreover we need to have R_x, C, I, as finite constants, such as $A(i,j;k,l)$ and $B(i,j;k,l)$, and this is an important key information, because there are more sophisticated models in which some of these assumptions are no longer valid.

In most of the applications it is important to guarantee the network stability. For example we examine the problem of the image processing: we have an image as the input of the network; the output image, after the CNN transient, is a modified version of the input. We can do it into two steps:

1) We fix the input voltages v_{uij} and the initial values of the state variables $v_{xij}(0)$ as grey levels of the image to process (we suppose to normalise the grey scale into the interval [-1,1]) and we associate each pixel of the image to the state of each network cell [2].

2) We make the network to evolve and, when the transient is finished, the system will go towards a stable equilibrium point and the outputs v_{yij} will give us the results of simulation that is the output image.

Just to give more information, we can say we are able to process a colour image, by having three CNNs, each of them used to process one of the fundamental colours, treating it just like a grey scaled image, like in a television set.

The problem of the CNN programmability

The most important problem we have with CNNs is that we don't have a unique method to assign the correct values to the cloning template coefficients for a specific application. However there are a large variety of libraries of templates, which are heuristically obtained.

Now we present some results, which are treated with more accuracy in [2] and [3], as regards image processing applications.

The first application we present is the noise removal in which we haven't the B template, but the array begins with the state capacitor already charged to the initial state. We present an array with a neighbouring radius r = 1.

We can see from the analysis of the *A* template the pixels of the image will take on a value that is the weighted average of its original intensity and its neighbours.

$$A = \begin{bmatrix} 0.0 & 1.0 & 0.0 \\ 1.0 & 2.0 & 1.0 \\ 0.0 & 1.0 & 0.0 \end{bmatrix}$$

A further example is given by the *edge extraction*, which is achieved using a template *A* with no couplings from the neighbouring outputs to the cell.

$$A = \begin{bmatrix} 0.0 & 0.0 & 0.0 \\ 0.0 & 1.0 & 0.0 \\ 0.0 & 0.0 & 0.0 \end{bmatrix}$$

$$B = \begin{bmatrix} -0.25 & -0.25 & -0.25 \\ -0.25 & 2.0 & -0.25 \\ -0.25 & -0.25 & -0.25 \end{bmatrix}$$

For this application the bias current has to be set to -0.25 μA.

The third and last application here presented do not cover the image processing purpose but arc rcfcrrcd to the solution of partial differential equations. Briefly a partial differential equation is an equation of the type:

$$\left(\frac{\partial^2}{\partial x^2} + \frac{\partial^2}{\partial y^2}\right).\vec{E} = \frac{1}{c^2}\frac{\partial^2 \vec{E}}{\partial t^2}$$

the wave equation, in which E is the module of the electric field. The heat diffusion equation is an example of partial differential equation:

$$\frac{\partial \rho}{\partial t} = \frac{\mathrm{K}^2}{\mathrm{C}^2}\left(\frac{\partial^2}{\partial x^2} + \frac{\partial^2}{\partial y^2}\right).\rho$$

where K is heat conductivity, C the heat capacity of the material and ρ the heat per unity area.

The feedback template for this application is the following:

$$A = \begin{bmatrix} 0.0 & 0.25 & 0.0 \\ 0.25 & 0.0 & 0.25 \\ 0.0 & 0.25 & 0.0 \end{bmatrix}$$

which corresponds to the first order approximation of the laplacian equation.

The CNN Integrated Design

A first consideration we have to point out is that in the discrete CNN circuit the voltages are referred to the input, the output and the state variables, while the currents are referred to the bias. But in the VLSI implementation we have to deal with currents because is this way the practical integrated implementation is easier. Moreover we utilised the ST Microelectronics technology HCMOS6 in which the library transistors MOS have a minimum channel length of 0,35 μm, with a 3,3 Volt single positive power supply.

According to these considerations it derives that we have to reinvent the practical implementation of the cell in a different manner with regards to the discrete implementation.

Taking into account earlier works developed in STMicroelectronics on similar arguments [4] as a starting point, we have first to consider the CNN equation (5.a), i. e. the sum of currents that flow together in a node. In fact we have the current that flows across the resistance R_x which is proportional to the state variable of the cell, the current across the capacitor C, the contributions of the voltage controlled current supply from the neighbouring cells and then the template coefficients which have to be implemented like some transconductance circuits (for example before the state template we must transform the state of other cells, which is a voltage, into a current), and the output current due to the non-linearity, which input is a voltage.

We have also to take in account that we operate with a single positive power supply and therefore the state and the non-linearity will never assume negative values, and hence we must re-scale all the equations.

Moreover one issue to be pointed out regards the measure of the coupling with the neighbouring cells. Since this is an experimental device we decided to implement the simplest, that is the $r = 1$, according to the chip requirements defined inside work that developed in the DICTAM european project.

The modified equation

We start from the equation of a 3x3 CNN in which the resistance R_x and the capacitor C are present, and we compare it with the normalised equation. In the most general case, in which are present the feedback template A, the control template B, and the state template C:

$$C\dot{x}_{ii} = -R^{-1}x_{ii} + \sum_{j=1}^{9} \hat{A}_{ij}y_{ij} + \sum_{j=1}^{9} \hat{B}_{ij}u_{ij} + \sum_{j=1}^{9} \hat{C}_{ij}x_{ij} + I \tag{7}$$

$$y_{ij} = \frac{1}{2}\left[\left|x_{ij} + 1\right| - \left|x_{ij} - 1\right|\right] \tag{8}$$

$$\dot{x}_{ii} = -x_{ii} + \sum_{j=1}^{9} \hat{A}_{ij} y_{ij} + \sum_{J=1}^{9} \hat{B}_{ij} u_{ij} + \sum_{j=1}^{9} \hat{C}_{ij} x_{ij} + I$$

We have to look at for the trans-characteristic of non-linearity as it can be implemented into an integrated chip using a single power supply, to get the first transformation in the equation. In fact, by drawing this characteristic we obtain the graph of Figure 4:

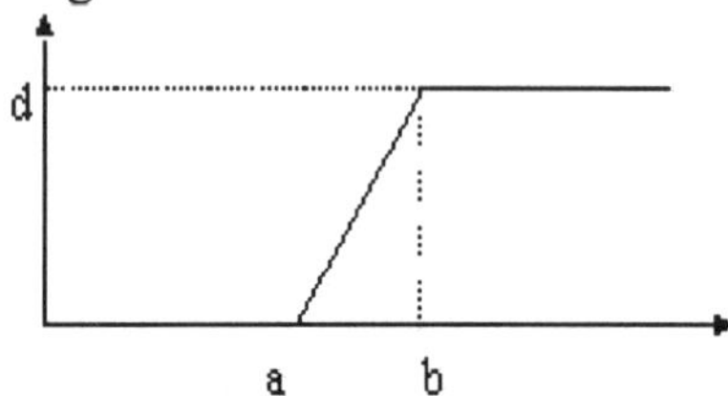

Figure 4: Trans-characteristic of the non-linearity

We may observe that the non-linearity doesn't start from a negative voltage value, and isn't centred on the zero. Therefore the first transformation is a linear one, by which we change the variable x into the variable x':

$$x' = Gx + h \quad \text{and} \quad G = \frac{b-a}{2} \quad e \quad h = \frac{b+a}{2} \tag{9}$$

Solved with respect to x', we have:

$$x = G^{-1} \cdot x' - G^{-1}h \quad \textit{and} \quad G^{-1} = \frac{2}{b-a} \tag{10}$$

By substituting the (10) in equation (8), we obtain:

$$y = \frac{1}{2}\left[\left|G^{-1}x' - G^{-1}h + 1\right| - \left|G^{-1}x' - G^{-1}h - 1\right|\right] = \tag{11}$$

$$= \frac{1}{2}G^{-1}\left[\left|x' + G - h\right| - \left|x' - G - h\right|\right] = G^{-1}y'$$

with the position we may have from Figure 4:

$$y' = \frac{1}{2}\left[\left|x' + G - h\right| - \left|x' - G - h\right|\right] = \tag{12}$$

$$= \frac{1}{2}\left[\left|x' - a\right| - \left|x' - b\right|\right]$$

Thus we have:

$$G^{-1}\dot{x}_{ii} = -G^{-1}x'_{ii} + G^{-1}\sum_{j=1}^{9}\hat{A}_{ij}y'_{ij} + \sum_{j=1}^{9}\hat{B}_{ij}u_{ij} + G^{-1}\sum_{1}^{9}\hat{C}_{ij}x'_{ij} + \\ -G^{-1}\sum_{1}^{9}\hat{C}_{ij}h + I + G^{-1}h \tag{13}$$

By dividing both of the members of (13) by G^{-1} we have:

$$y' = \frac{1}{2}\left[\left|x' - a\right| - \left|x' - b\right|\right] \\ \dot{x}'_{ii} = -x'_{ii} + \sum_{J=1}^{9}\hat{A}_{ij}y'_{ij} + G\sum_{J=1}^{9}\hat{B}_{ij}u_{ij} + \\ + \sum_{j=1}^{9}\hat{C}_{ij}x'_{ij} - \sum_{J=1}^{9}\hat{C}_{ij}h + GI + h \tag{14}$$

Now we have to introduce x'_{ii} in the sum of the $\hat{C}_{ij}$ terms, and we impose a translation of y' in such a way to translate the characteristic above the horizontal-axis. Thus we obtain:

$$y'' = \frac{d}{2}(y' + 1) \\ \dot{x}_{ii} = \sum_{J=1}^{9}\hat{A}_{ij}y'_{ij} + G\sum_{J=1}^{9}\hat{B}_{ij}u_{ij} + \sum_{J=1}^{9}\left[\hat{C}_{ij} - \delta_{ij}\right]x'_{ij} + \\ -\sum_{1}^{9}\left[\left(\hat{C}_{ij} - \delta_{ij}\right)h\right] + GI \tag{15}$$

Now, by getting y' out of y'' we'll have $y' = \frac{2}{d}y'' - 1$, and by substituting it in the equation (15):

$$\dot{x}'_{ii} = \frac{2}{d}\sum_{J=1}^{9}\hat{A}_{ij}y''_{ij} - \sum_{J=1}^{9}\hat{A}_{ij} + G\sum_{J=1}^{9}\hat{B}_{ij}u_{ij} + \sum_{J=1}^{9}\left[\hat{C}_{ij} - \delta_{ij}\right]x'_{ij} + \\ -\sum_{J=1}^{9}\left[\left(\hat{C}_{ij} - \delta_{ij}\right)h\right] + GI \tag{16}$$

Now, let us consider the transformation as it can be shown by the following Figure:

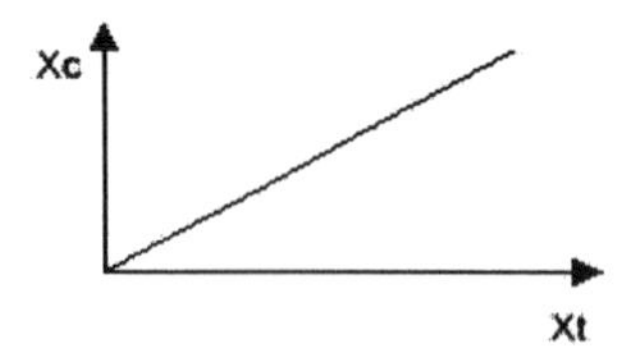

Figure 5: Conversion voltage to current of the state of the cell

The voltages, which represent the state variables of each cell of the neighbourhood, are converted into currents and therefore analytically we have a new transformation of the quantity x' into the quantity x'', inside the state template and hence:

$$x'' = \gamma x'$$

$$x'_{ii} = \frac{2}{d}\sum_{J=1}^{9}\hat{A}_{ij}y''_{ij} + G\sum_{j=1}^{9}\hat{B}_{ij}u_{ij} + \tag{17}$$

$$+\sum_{J=1}^{9}\left(\left[\hat{C}_{ij} - \delta_{ij}\right]\frac{1}{\gamma}x''_{ij}\right) - \sum_{J=1}^{9}\hat{A}_{ij} - \sum_{J=1}^{9}\left[\left(\hat{C}_{ij} - \delta_{ij}\right)h\right] + GI$$

in which we have denoted with γ the slope of the line of Figure 5. Thus we obtain:

$$I' = GI + h - \sum_{J=1}^{9}\hat{A}_{ij} - \frac{1}{\gamma}\sum_{J=1, j\neq i}^{9}\hat{C}_{ij}h$$

$$x'_{ii} = \frac{2}{d}\sum_{1}^{9}\hat{A}_{ij}y''_{ij} + G\sum_{J=1}^{9}\hat{B}_{ij}u_{ij} + \frac{1}{\gamma}\sum_{J=1}^{9}\left(\left[\hat{C}_{ij} - \delta_{ij}\right]x''_{ij}\right) + I' \tag{18}$$

And thus:

$$\frac{d}{2}x'_{ii} = -\frac{d}{2}x'_{ii} + \sum_{J=1}^{9}\hat{A}_{ij}y''_{ij} + \frac{d}{2}G\sum_{J=1}^{9}\hat{B}_{ij}u_{ij} + \frac{d}{2\gamma}\sum_{1}^{9}\left(\hat{C}_{ij}x''_{ij}\right) + \frac{d}{2}I' \tag{19}$$

By using the following positions:

$$
\begin{aligned}
R &= \frac{2}{d} && \Omega \\
C &= \frac{d}{2} && \mathrm{F} \\
I'' &= \frac{d}{2} I' && \mathrm{A} \\
\hat{B}'_{ij} &= \frac{dG}{2} \hat{B}_{ij} \\
\hat{C}'_{ij} &= \frac{d}{2\gamma} \hat{C}_{ij} && \Omega^{-1}
\end{aligned}
\tag{20}
$$

we may get the following equation:

$$
C\dot{x}'_{ii} = -\frac{x'_{ii}}{R} + \sum_{J=1}^{9} \hat{A}_{ij} y''_{ij} + \sum_{J=1}^{9} \hat{B}'_{ij} u_{ij} + \sum_{J=1}^{9} \hat{C}'_{ij} x''_{ij} + I'' \tag{21}
$$

In such a way we arrive to this equation that is close to the canonical CNN equation.

The cell implementation

First of all we may see that the non-linearity implementation is one of the most important part of the circuit because it is responsible for the cell behaviour.

Furthermore, we must think to the correct programmability of the cell, which is given by the template coefficients and the bias, and we will see the way to do this. Moreover, according to the transformation of Figure 5 we need for a rail to rail voltage to current converter.

Finally we must consider the problem of implementation of a sum circuit where all the currents flow.

The non-linearity

In Figure 6 we may see the non-linearity scheme:

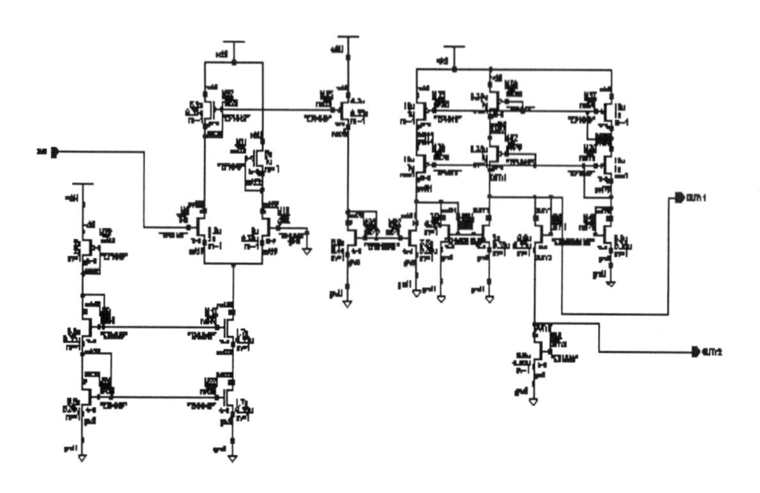

Figure 6: Schematic of the non-linearity

The implementation of non-linearity is quite simple, because it takes advantage of the MOS voltage to current characteristic. The output of the differential stadium gives a current that is proportional to the voltage applied since the saturation occurs. The linear zone will approach quite well the linear region of the non-linearity, while to have a more clear-cut profile of the region in which the output is constant we use a P-MOS mirror to fix the current to 20 μA. Substantially the operation we make is to subtract this current to the current we bring by a transistor acting a mirror from the input differential stage. This difference will be a current whose initial value is 20 μA, decreasing linearly to 0 A, and therefore its behaviour resembles the non-linearity but in a reflected way, and thus we have to reflect it with an identical subtraction circuit, and the result is our non-linearity. We are able to shift the zero point of the resulting current modifying the shape ratio of the transistors of the input differential stadium increasing the right one and decreasing the left one. We are also able to module the range of input voltages for which the current is linear by changing the shape ratio of the MOS diodes acting the subtraction. The result we have is 1.05 V to 1,7 V of linearity range, an interesting range possible thanks to the limited number of utilised components. The two output cascode MOS are essential, because the non-linearity is connected to the digital multipliers that we'll see in the following, and these circuits operate with cascode mirrors.

Moreover it's possible to design this circuit tolerating enough well inaccuracies because of the robustness of this kind of structure, but in simulation the maximum current value was 19,92 μA, very close to the desired value. In Figure 7 we may see the results of the simulations we made:

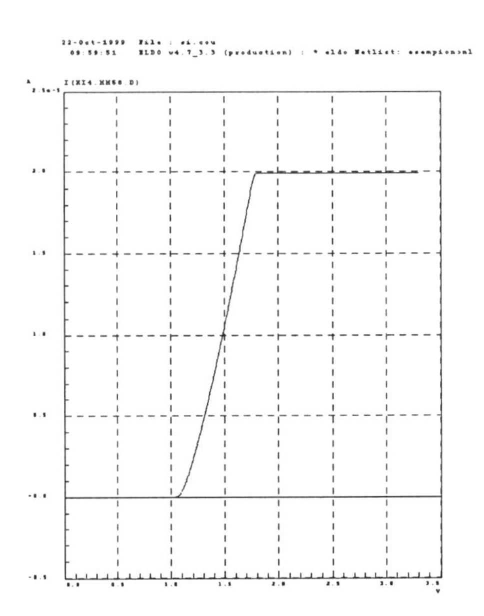

Figure 7: Results of simulation of non-linearity circuit

The template coefficients

The template coefficients are the components that play the most important role of the circuit operating, because they determine the cells programmability. In fact, all the complexity of such a structure lies in the digital circuitry, that interfaces with the analog one (the cells). [5], [6], [7]. In such a circuit the values given to the templates are digital number of 7 bits, which are converted into analog currents, and this is the same thing we do in our structure, by means of the digital multipliers.

In [6] and [7] we may see that the major applications of such circuits require a neighbourhood radius 1, but they can operate with an higher connectivity degree while our circuit, which is an experimental one, is simpler and cannot vary its connectivity.

However the next step of such a work could be to realise a chip with the possibility to program the coupling degree among the cells and to have, therefore a more dynamic and powerful tool for elaborating input data. Hence, the first thing to consider is the number of bits to store the information, and we decided to use 8 bits, the first being the bit of the sign, so that we may store negative values, according to the applications.

The second thing to consider is that all the coefficients have values belonging to the same range, and this simplifies our work, because we may utilise an identical model. As we may deduce by (15) and (16) we may include in the state template the coefficients of x_{ii} so that one of the multipliers will have a bias term while the others may start from the zero.

As already mentioned, it was difficult to give the proper values to the coefficients of the cloning templates by using an algorithm, and the values that are provided, are heuristically provided.

By inspecting all the existing library template coefficient values we may have an idea of the variation range of these numbers. It can be noticed that the most recurrent values are 0.25, just like 0.5 and 1. Moreover the coefficients may be negative ones, and this is the reason for the sign bit.

Finally, we have a wide range of the value variations: from zero to six, and for the most common applications provided from a 3x3 CNN array we have: 0.0, 0.07, 0.1, 0.25, 0.5, 1.0, 1.5, 2.0, 2.5, 3.0, 4.0, 5.0, 6.0.

Figure 8 shows the schematic of a digital multiplier referred to the feedback template, that is the same for the other templates.

Substantially the circuit is a set of mirrors, in which the different values of the shape ratio determine different output current. Every branch is composed of three MOS, the two cascode realising the multiplying factor while the other acting as a switch, controlled by the line that carries the digital information. Since we have N-MOS like a switch, when the input is high, the transistor is on and the information is transmitted. When the transistor of the sign minus is on the current is mirrored through the cascode mirror and therefore the output current will have an opposite direction.

The output mirror is composed of transistors whose shape ratio has such dimensions as to sustain the maximum current of the multiplier, when all the transistors are on.

All the mirrors of the multipliers have shape ratio that differs each other by a power of 2, so that we'll have this value: 2, 4, 8, 16, 32, 64 and 128, so that the current flowing in the different branches ranges form 20 μA to 128μA. To cover all the range of values with seven bits we gave a step of 0,025 and this step is reflected in the shape ratio of the MOS M4 and M154 of the P-MOS mirror on the left of Figure 8.

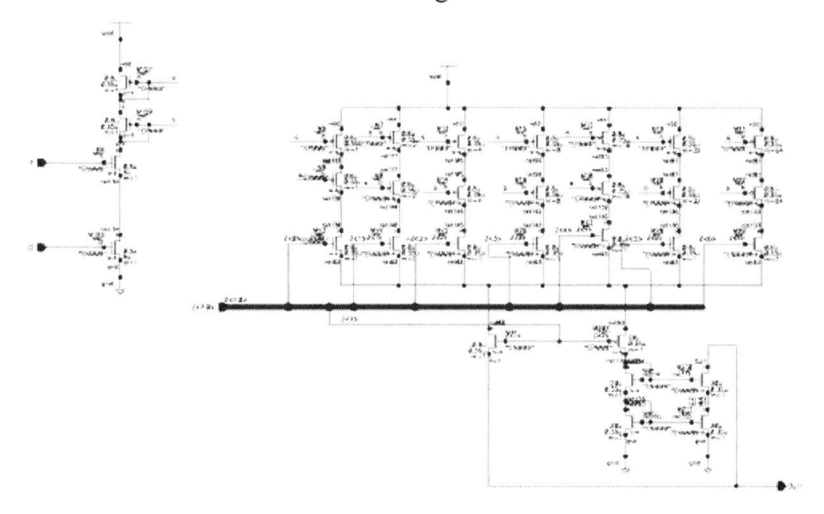

Figure 8: Schematic of the basic template's coefficient

As we may see from the figure the two input transistors of minimum ratio which make the right side of a mirror, carry the information of the other share of the cell.

We have chosen P-MOS transistors to weight the input, so we have the output current direction which leaves the circuit while if the sign bit is V_{DD} the current direction is entering the circuit.

Furthermore we have to say that the cascode structure is important to correctly implement the multiplier because of the intrinsic non linearity of the MOS transistor, and the cascode configuration avoid the deviations from ideal behaviour when the inputs are high, and this is the reason for which the output non-linearity is a cascode configuration. In Figure 9 the results of simulation of such a circuit are shown.

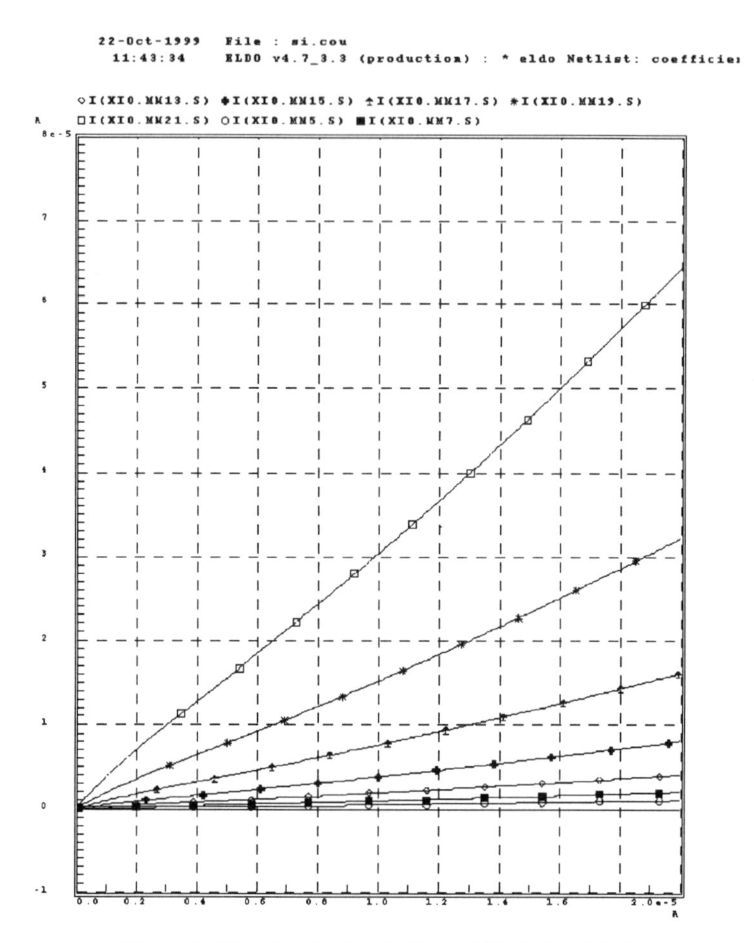

Figure 9: The circuit simulation of digital multiplier

The bias

The bias is basically an independent current supply, that is digitally programmable and therefore its structure is a digital multiplier. According to (18) and (20) the bias is the sum of many contributions such as the feedback template coefficients and the state template coefficients, and the intrinsic share I, hence the bias term is realised with the input like a sum circuit. Furthermore we have to re-scale the shape ratio of the input MOS in order to have the desired value we obtained from the previous equations.

Figure 10 shows the basic circuit of bias, in which we may note the seven bit multiplier, and it's valid only for positive value of current whose direction is leaving the circuit.

There is also the negative portion of the circuit for the current entering the circuit. The 8th current mirror is the constant term, as we may obtain from the above equations.

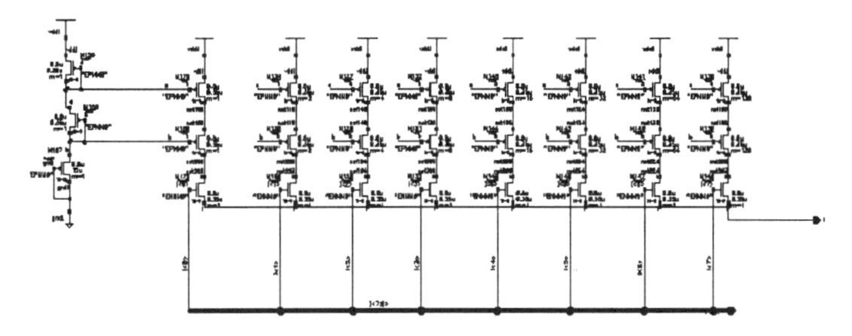

Figure 10: Schematic of bias circuit

We have to realise two different bias circuits, because of the constant term implementation, that we realised either with P-MOS and N-MOS.

The output of such a circuit is referred to the line, and it either will take away or let in the current. In Figure 11 we show the circuit of the sum, which is a set of mirrors to have on output the sum of all the terms.

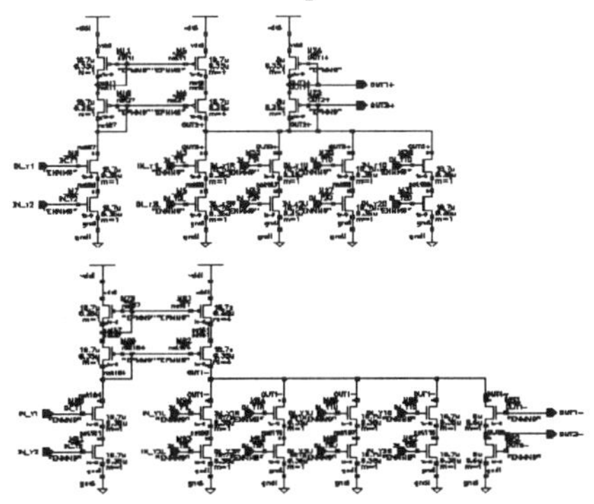

Figure 11: Schematic of sum circuit

The voltage to current converter

This component is a very important one, since the state, which is a voltage has to be converted into a current to be manipulated by the digital multiplier.

According to the transformation introduced in Figure 5, this conversion has to be highly linear over all the range of input voltage. Therefore the converter is to be a rail to rail converter, because the transconductance amplifier has a very high distortion due to non-linearity of MOS transistor.

The solution to this problem was provided by an additional transconductance amplifier which implements the inverse function realised by the first converter. The result is known as the cross-coupled differential pair, and the aim of such a configuration is to annihilate the effects of the higher harmonics. In such a way we want to have a resulting current of the order of 10 μA.

The other components

The first of the components we consider is the capacitor, which is a technology capacitor that is present in the hcmos6 library, just like the resistance. Although the unavoidable tolerances and the parasitic effects the structure is operating thanks to the robustness of CNN circuits. The value of the capacitor is a few fF, and the resistance of a few kΩ so that the resulting time constant of the cell will be very short and the speed of elaboration very enhanced.

Finally we present the schematic of a CNN cell in Figure 12 as a result of this work:

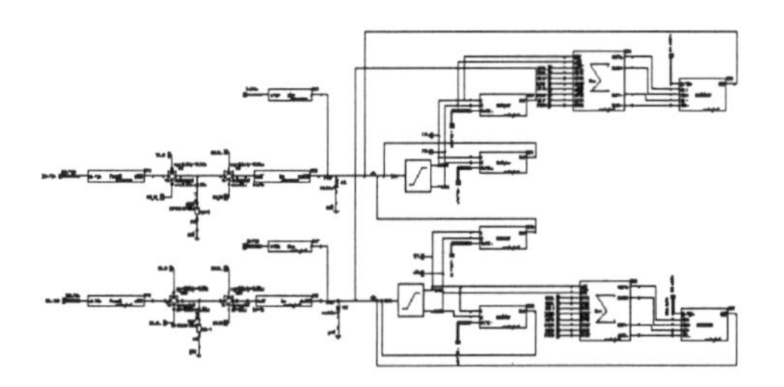

Figure 12: Schematic of the CNN cell

Conclusions

We have shown the possibility to make an integrated chip for realising a CNN architecture and we saw how to implement the basic components of a cell. One of the most important expects that come out is the robustness of this structure. In fact by the ELDO temperature variation simulations we observed that the basic operations were guaranteed, notwithstanding the strong variations of some of the parameters, since they compensate each other.

Moreover this architecture is a very simple one, because all the parts are realised with well known structure, and the repetition of these structures that is the consequence of the nature of CNN allows to easy integrate such a circuit.

We said the most relevant problem of CNN is the programmability, that is the way we may use such a circuit. The only way we may get over this difficulty is to think about some specified applications, so that, by programming the chip with few values, we may have a powerful tool for parallel computation.

We have to say that, obviously this chip, besides the analog circuitry of CNN, have to be composed by a digital circuitry which is the interface with the templates coefficients. In this part of circuit will arrive the information, that is the eight bits. In the DICTAM european project there are portions of used circuit to store these information. They are composed of registers with 16 bits, since they store information for two coefficients at a time [7].

Therefore the future of CNN classic architecture is in these well defined and specific applications, in fact, for example, it seems that FIAT wants to apply CNNs in the so called "Smart rear-view mirrors", that are the driver-mirrors, which are oriented themselves in such a way to allow driver to better watch all the obstacles.

The other application is a chip that replaces the retina and interfaces itself with the optic nerve to take care of some blindness form. There is a new of some months ago, that in Germany some scientists realised such a circuit with a neural net architecture, and therefore at present the study in this sector is at the beginning.

References

[1] Leon O. Chua and Lin Yang, *Cellular Neural Networks: Theory and Applications* Ieee Transactions On Circuits And Systems, Vol. 35, N° 10, October 1988.
[2] Leon O. Chua and Lin Yang, *Cellular Neural Networks: Applications*, Ieee Transactions On Circuits And Systems, Vol. 35, N° 10, October 1988.
[3] Leon O. Chua, Lin Yang, K. R. Krieg, *Signal Processing Using Cellular Neural Networks*, Journal Of Vlsi Signal Processing, 3, 25-51 (1991).

[4] Francesco Doddo, *Architettura Monolitica Di Rete Neurale A Doppio Strato Per Il Controllo Di Sistemi Multiattuatore*, Tesi di Laurea- Anno Accademico 1997-98 Facoltà di Ingegneria, Università degli Studi di Messina.
[5] G.Linan, R.Dominguez-Castro, S.Espejo, A.Rodriguez-Vasquez, *Cnnuc3 User Guide-Preliminary Version*, Centro Nacional De Microelectronica.
[6] E.Roca, S.Espejo, R.Dominguez-Castro, A. Rodriguez-Vasquez, *Imagav-1: Technical Report- Preliminary Version*, Institute of Microelectronic of Seville.
[7] G. Linan, S. Espejo, R. Dominguez-Castro, E. Roca, A.Rodriguez Vasquez, *Cnnuc3: A Mixed-Signal 64x64 Cnn Universal Chip*, Institute of Microelectronic of Seville.

Design of a Switching Regulator with Fuzzy Control in BCD Technology

Giuseppe Scalia

Accent S.r.l.

Via F. Gorgone 6, 95030 Catania - Italy

E-mail: giuseppe.scalia@accent.it

Introduction

An intense activity of research it does exist today that aims to cover important sectors in which is relevant the use of integrated DC/DC converters with elevated performances like high answer to the load transitory with wide range of stability.

We think, for instance, to the DC/DC converters used in the mobile phone, in the hard disk, and in portable apparatuses generally. In these last years the electronics factory has produced and commercialized numerous integrated devices type general purpose DC/DC converter.

However, the increase of demands has required the customizzation of the final product oriented to satisfy high target. The idea of *'client satisfaction'* has push us toward the present job, and in collaboration with the Switching-Regulators Power-Conversion group of the ST-Microelectronics in Catania, has been developed a control system based on a fuzzy algorithm that can be easily integrated allowing the realization of a switching regulator with fuzzy control leading all benefits associate to this new control methodology. The used technology is the BCD one.

The converter object of this study is known in literature as *step up (boost) converter* at fixed frequency with duty cycle control (PWM controller). In the major of cases the control of the duty cycle is realized through a simple scheme of current mode linear regulation. The limited flexibility of this control technique in the step up continuous mode DC/DC converters, pushes us to seek a better solution, within the non linear control techniques, like that innovative obtainable through the introduction of a fuzzy algorithm in the voltage loop of a traditional architecture.

Principle of operation

The voltage regulators are circuits that, absorbing input power from a generator of not regulated voltage, make it available to the output with a stable value independently from any variation of the input and the load. Two are the most common families of regulators: the series linear regulators and the SWITCHING regulators. The switching regulator has a wide control possibility, reaching elevated performance (80%—90%) and allow the over-voltage (*step-up configuration)* of the input voltage.

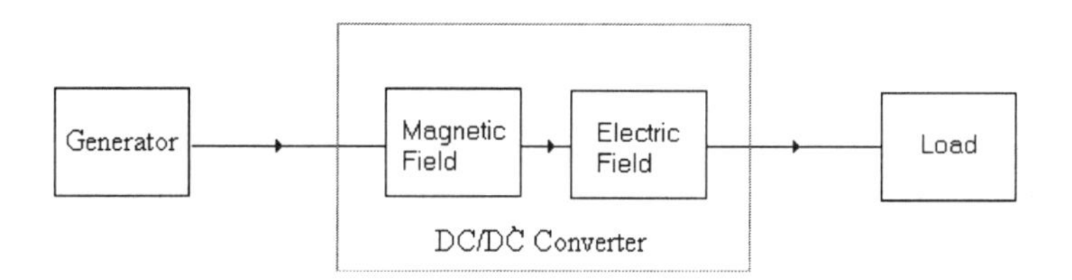

Figure 1: Energy Flow *"power-load*" regulated by a DC/DC converter

The DC/DCs converters are a key the transfer of energy between the generator and the load, storing in a first phase, energy in the magnetic field of an inductor that is then transfered in the electric field of a condenser in parallel to the load as shown in Figure 1.

In Figure 2 the *Step-Up* topologies of DC/DC converters is shown.

The two passive elements are recognized: inductor and condenser, and the two power switch are respectively schematized, with a bipolar transistor (SW1) and with a diode (SW2). The charge of the inductor and the subsequent energy transfer to the condenser are regulated by an opportune timing signal (clock), to the purpose to maintain constant the output voltage respect the variation of the absorbed load current.

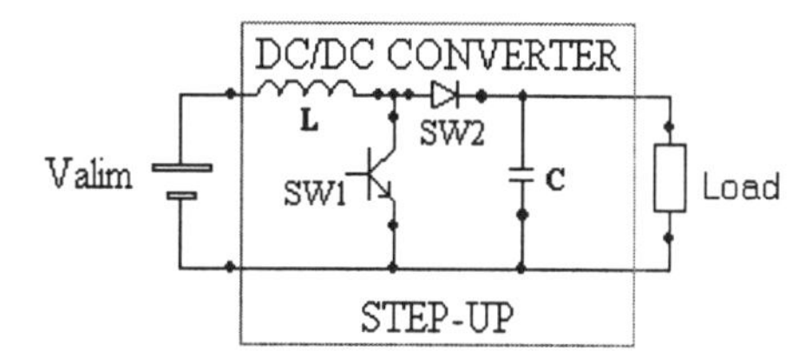

Figure 2: Circuit scheme of the Step-Up topology

In every commutation cycle of duration T, in general, three intervals of consecutive time, characterized by the switches state can be individualized:

T_{ON} (transistor-on time): the transistor *SW1* is ON and the diode *SW2* is OFF. In this interval energy is stored in the magnetic field of the inductor;

T_D (diode-on time): the transistor *SW1* is interdicted by the control signal while the diode *SW2* is ON; in this phase the energy stored in the inductor decrease, due to the transfer of energy from the magnetic field to the electric field of the capacitance;

T_I (idle time or free-wheeling time): *SW1* and *SW2* are both OFF, all the energy required to the load is furnished by the capacitance, which decreases the voltage to its terminals, while any current flow in the inductor and its magnetic field results null.

The sum of the three time intervals is equal to the total period of commutation: $T = T_{ON} + T_D + T_I$.

The presence of T_I determines the two possible operation mode for the DC/DC converters:

Continuous mode (CCM) in which the current in the inductor, and therefore its magnetic field, is never null in the whole cycle $(T_I = 0)$.

Discontinuous mode (DCM) in which for an interval of time the current in the inductor is annulled $(T_I \neq 0)$.

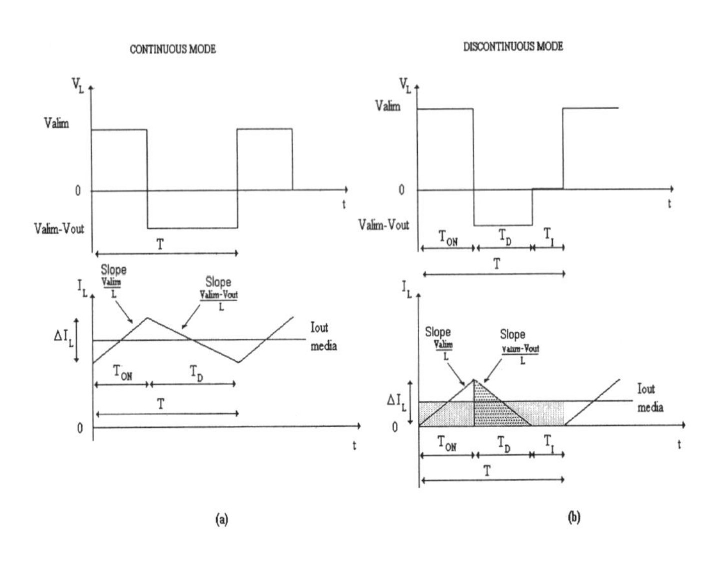

Figure 3: Step-Up: current and voltage wave form in CCM (a), DCM (b).

Comparison between CCM and DCM mode in the Step-Up converter

The Step-Up converters in CCM mode are characterized by a control law that conduct the system, under particular load condition, versus the instability, characterized by strong output ripple and sub-armonich oscillations [3].

Opposite, the DCM Step-Up converters have good stability performance but in comparison to the load current they are characterized by more peaks of the maximum currents which flow through the switches and the reactive components. The better stability system of the DCM makes it more common and advantageous in comparison to the CCM but the consumption problem, the dimensions of the components in the systems powered by batteries, push more and more the technique to develop highly sophisticated controller for Step-Up converters able to work without problems of instability in two CCM and DCM mode.

Objective of this job is, in fact, the project and the integrated realization of one of such controllers.

Techniques of Control for DC/DC Converters

PWM control: voltage-mode and current-mode

A DC/DC converter has the purpose to furnish energy to a load, supplying constant voltage despite variations in power voltage and power current of the and/of the current absorbed by the load. To this purpose it is necessary to use a closed loop control system, that produces a logical signal which acts on the activation of the power switch SW1 (see Figure 2) dosing the flow of power transferred to the load. This control, if effected modulating the duration T_{ON} of the switch on of the power switch, in comparison to the duration T of the signal to a fixed frequency $f_s = \frac{1}{T}$ that regulates the switch on, is said PWM, that is control of the duty-cycle $d_1 = \frac{T_{ON}}{T} = T_{ON} \cdot f_s$.

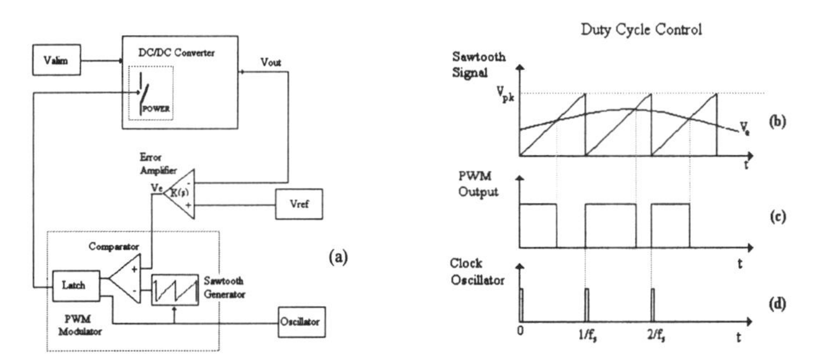

Figure 4: Blocks scheme of a PWM controller and main shape wave

In Figure 4-a the simplified blocks scheme to a PWM controller in his general form is shown. It is possible to distinguish: an oscillator that determines the time frequency f_s of the switch on of the power switch and an error amplifier that produces a signal V_e given by:

$$V_e = K \cdot \left(V_{REF} - V_{OUT}\right) \tag{1}$$

were K is the static gain of the amplifier, V_{OUT} the regulated output voltage and V_{REF} the reference voltage inside the regulator.

The blocks scheme is composed, besides, by a comparator that effects the comparison between the error signal V_e and a periodic signal of period $\frac{1}{f_s}$ (see Figure 4-b). The output of the comparator is applied, together to the clock signal (Figure 4-d), to a latch whose output (see Figure 4-c) determines the switch on and the turn off of the power switch.

If the constant signal is always given with the same $\frac{dV}{dt} = const$, for example using the oscillator that gives the clock, the PWM control will be said **Voltage-mode**.

If rather the signal has a slope $\frac{dV}{dt} = \alpha \cdot \frac{dI_L}{dt}$ where α is the constant of proportionality and I_L current in the inductor, the control PWM is said **Current-mode** [2].

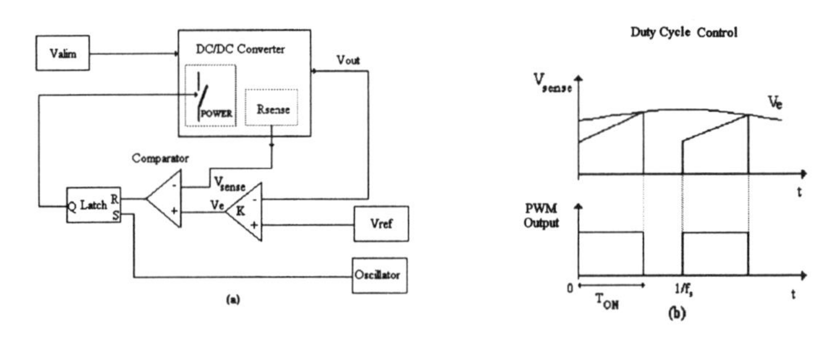

Figure 5: Blocks scheme of a PWM Current-Mode controller and main shape wave

Figure 5-a shows the blocks scheme of a regulator with PWM Current-Mode control, in which two rings of feedback are recognised. In the outer ring take place the control of the output voltage, in the inner one the information relative the current of the inductor [1] is used [4]. This type of regulators uses a sawtooth signal $V_{sen\,se}$ proportional to the current I_L that crosses the inductor:

$$V_{\text{sen}\,se} = \beta \cdot I_L \cdot R_{\text{sen}\,se} \tag{2}$$

where β is a constant of proportionality, $R_{\text{sen}\,se}$ a small value resistance in series to the switch and crossed by the inductor current I_L that flows during the interval T_{ON}.

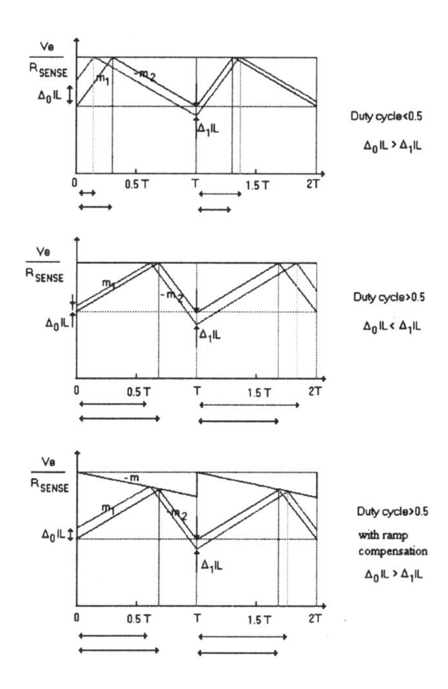

Figure 6: Instability for a PWM "current mode" converter

A technique of current-mode control, frequently used, is said Peak Current mode [2].

In this technique the power switch is turned off when the voltage signal proportional to the peak current of the inductor $\mathrm{V}_{\text{sen se}}$ reaches the value of the signal V_e in output of the error amplifier like in Figure 5-b: in such way it is able to have a faster answer to the input voltage and current output variations with an intrinsic limitation to the peak current against the output overloads.

One of the disadvantages that the Current-mode control introduces is the intrinsic instability of the closed loop system for values $d_1 > 0.5$ that is manifest through annoying sub-harmonic oscillation of V_{OUT} and perturbations of the current. This limitation can be overcome subtracting a ramp of voltage said of *compensation* with constant inclination *m* to the error voltage V_e. This technique is note in literature as *Slope Compensation* [2] [5].

The pure Current-mode control $\Delta_0 I_L$ responds to a perturbation of the inductor current in the subsequent clock interval causing a greater perturbation $\Delta_1 I_L$ respect $\Delta_0 I_L$ if $d_1 > 0.5$ (Figure 6B) or smaller if $d_1 < 0.5$ (Figure 6-a). The relationship that ties the variations in two consecutive clock periods, results:

$$\Delta_1 I_L = -\Delta_0 I_L \cdot \frac{m_2}{m_1}. \tag{3}$$

In absence of *ramp of compensation*, the condition $|m_1| > |m_2|$ is equivalent to $d_1 < 0.5$. The value 0.5 guarantees intrinsic stability.

In the case $d_1 > 0.5$ the stability of the system can be assured subtracting to a V_e a voltage ramp signal of angular coefficient m , as in Figure 6-c. In such way, condition (3) is changed in:

$$\Delta_1 I_L = \Delta_0 I_L \cdot \left(\frac{-m_2 + m}{m_1 + m} \right) \tag{4}$$

or

$$m > \frac{m_2 - m_1}{2} \tag{5}$$

In the case of a Step-Up regulator in the most critical situation that happens for $d_1 = d_{1_{\max}}$ the (5) becomes [1]:

$$m_{\min} = \frac{V_{ALIM} \cdot (2d_{1_{\max}} - 1)}{2L \cdot (1 - d_{1_{\max}})} \tag{6}$$

Therefore by choosing a ramp of compensation with $m \geq m_{\min}$, for every cycle the previous perturbation bringing the system to the stability is reduced. A disadvantage that the Current-mode control technique introduces is the extreme sensibility to the noise: in fact the voltage $R_{sen\,se} \cdot I_L$ used to generate the sawtooth signal, $V_{sen\,se}$, is very small compared to the level of voltage V_e, therefore a small noise is able to modify the instant of the turning off of the power switch (Jitter-noise).

Such problems limit strongly the practical applications of the step-up converters in continuous mode operation.

The restricted possibility to control this type of systems with linear control techniques pushes to seek a better solution, mostly flexible to the variations of the contour conditions of the system, within the non-linear control techniques.

Amoung the possible approaches of non linear control, particular interest has aroused the fuzzy logic, justified by the simplicity of the controller synthesis, based on the transfer of the phenomenon knowledge in a set of linguistic rules describing the control in the different operative conditions.

Synthesis of Control System Based on Fuzzy Logic

The idea of the developed work is to integrate the classical *ramp compensation* current-mode control with a fuzzy logic control that acts on the transfer function of the system with the purpose to get a non linear regulation to compensate the negative effects of load variations.

The purpose is to join the advantages of the current mode control, with the flexibility of a fuzzy control system in the voltage loop, widening the range of stability in comparison to the traditional control. The reference schematic adopted is visible in Figure 7.

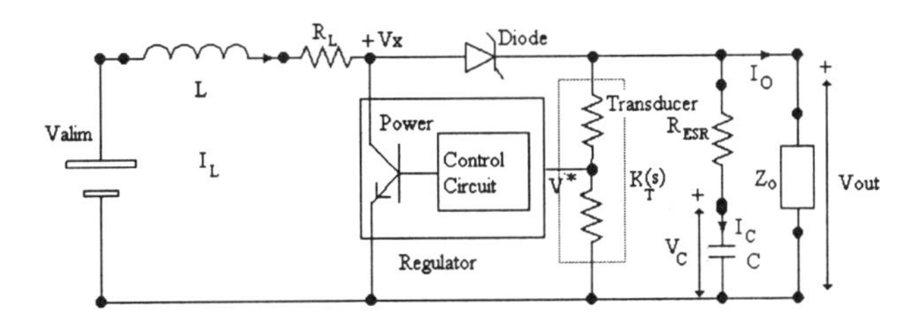

Figure 7: Reference schematic

The error and its first derived (D_E) are assumed as input to the fuzzy algorithm.

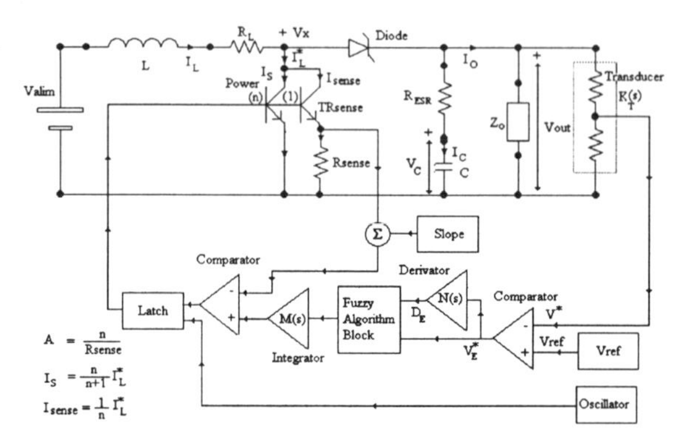

Figure 8: Schematic of the adopted control

The *error amplifier block* of the linear regulator had been replaced with a fuzzy control block wich inputs are the error and the first derivative of the output voltage. The output of such block is applied to one of the two input of the comparator that realises the PWM modulator and that governs the *switch off* of the power transistor. The aim is to act on the transfer function of the converter in non linear mode by compensating the negative effects of the linear control according to the principle scheme visible in Figure 8.

For the fuzzy algorithm, the linguistic rules are empirically derivated based on the following criterions:

- for a small error in absolute value, the output of the fuzzy block (*'F_Out'*) is determined in reason of the load condition (*'Derived'*);
- an elevated and positive value of the error forces *F_Out* to its maximum value;
- an elevated and negative value of the error forces *F_Out* to the least value.

Since the used frequencies don't allow the use of digital systems and/or fuzzy tools, an analogic fuzzy circuit will be designed and will be simulated in ELDO environment determining the parameters to realize a good control.

To the fuzzy algorithm implementation we used the BCD3 technology, and the gotten control has been simulated on a current mode PWM DC/DC regulator operating at 220 KHz.

The fuzzy algorithm

We define two linguistic variables in input to the fuzzy algorithm, that are:

Error (E): expression of the error (V_E) in the output voltage in comparison to the nominal voltage (V_{NOM})

Derivative (DE): expression of "error derivative" (DE) of the output voltage.

The aim of the fuzzy algorithm is to determine the value with whom compare, in the comparator that realize the PWM (Figure 8), the ramp proportional to the inductor current. A sufficient positive value of V_E ($V_{OUT} << V_{NOM}$) must determine the maximum output from the fuzzy block; indeed, an enough negative value of V_E ($V_{OUT} >> V_{NOM}$) must determine the least output.

Therefore, will be fixed three linguistic values for the *Error* variable:

Error neg
Error zero
Error pos

And for the *derivative variable* are individualized five linguistic values:

Very Small (VS)
Small (S)
Medium (M)
Big (B)
Very Big (VB)

Through a *'trial and error'* procedure we made the calibration of the memberships.

For the output variable, *F_Out*, the following seven linguistic values are individualized, each one associate to a rule:

ENeg for convenience *U[R1]*;
EZ_DeVS for convenience *U[R2]* (Error Zero_Derivative Very Small);
EZ_DeS for convenience *U[R3]* (Error Zero_Derivative Small);
EZ_DeM for convenience *U[R4]* (Error Zero_Derivative Medium);
EZ_DeB for convenience *U[R₅]* (Error Zero_Derivative Big);
EZ_DeVB for convenience *U[R6]* (Error Zero_Derivative Very Big);
EPos for convenience *U[R7]*.

The rules are defined so that is realized the necessary *gain loop modulation* to prevent the instability of the system.

The seven rules are the following:

R1: IF (*Error IS Error pos*) THEN *F_Out IS Epos*;
R2: IF (*Error IS Error zero*) AND (*Derivative IS VS*) THEN *F_Out IS EZ_DeVS*;
R3: IF (*Error IS Error zero*) AND (*Derivative IS S*) THEN *F_Out IS EZ_DeP*;
R4: IF (*Error IS Error zero*) AND (*Derivative IS M*) THEN *F_Out IS EZ_DeM*;
R5: IF (*Error IS Error zero*) AND (*Derivative IS B*) THEN *F_Out IS EZ_DeB*;
R6: IF (*Error IS Error zero*) AND (*Derivative IS VB*) THEN *F_Out IS EZ_DeVB*;
R7: IF (*Error IS Error neg.*) THEN *F_Out IS ENeg.*

Despite in the *antecedents* only the connective *AND* is present, the *inference* of each rule can be realized with whatever device able to pass the least between two fuzzy input greatness.

The *de-fuzzification* is gotten through the sum, weighed with the values of the *consequent* of the *activation degrees* for every rule.

In Figure 9 the values assumed for *F_Out* in the *consequent* are indicated, with reference to the rules, by the terms $U[R_I]$.

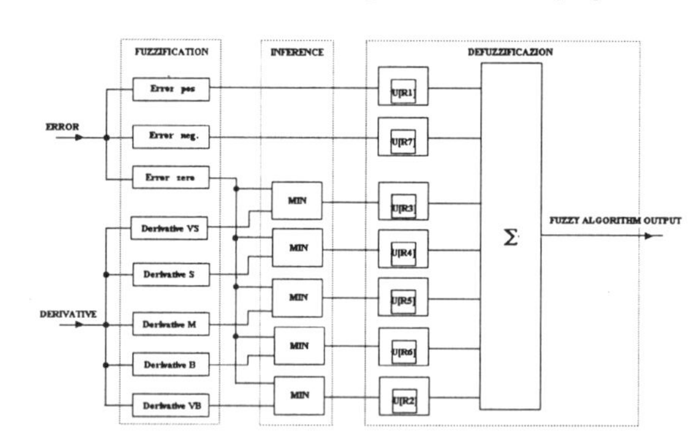

Figure 9: Scheme of the employed fuzzy algorithm

By implementing the early principle schematic, we obtain the fuzzy algorithm showed in the following page.

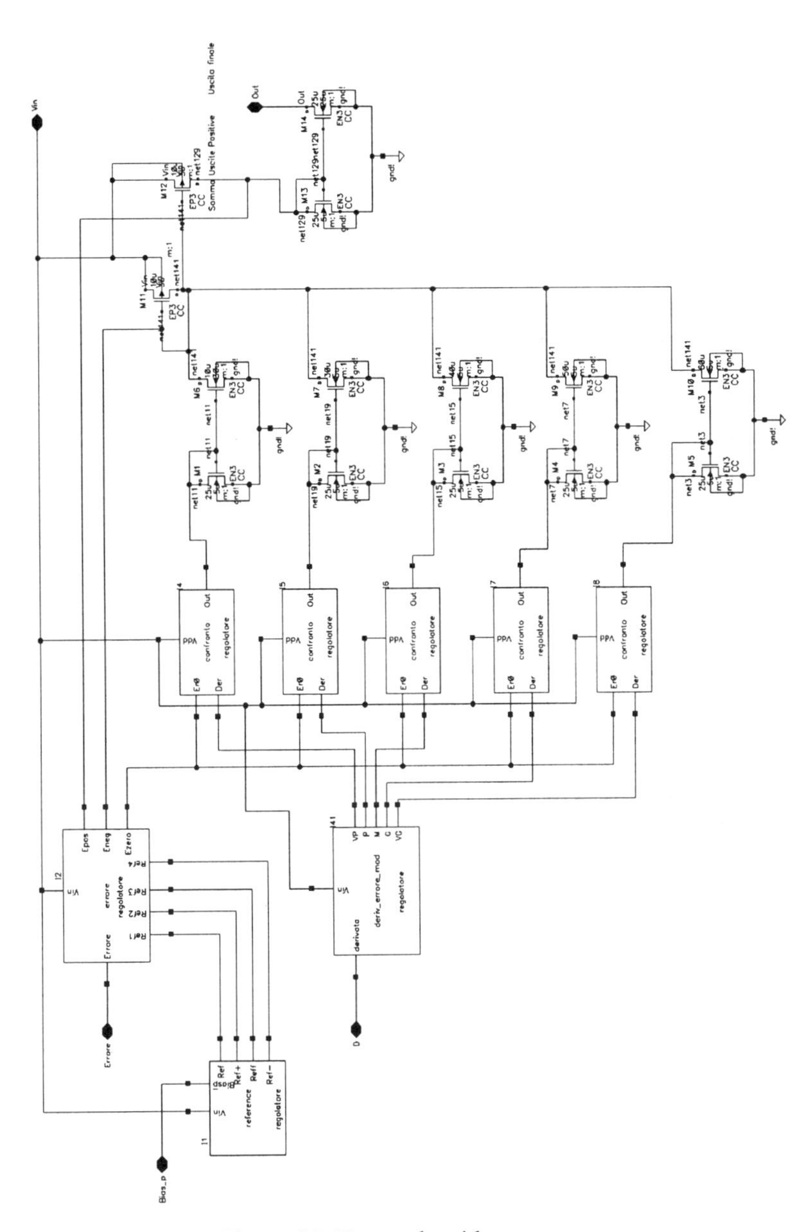

Figure 10: Fuzzy algorithm

In this scheme it is possible to distinguish the different functional blocks that characterize it. Inside the blocks *error* and *derivative error* we have the analog circuits that realize the memberships function, while the block "*confronto*" implements the circuit thet searches the *min* between the functions during the *de-fuzzification* step.

Simulation of the system with fuzzy control

With the purpose to verify the functional validity of the fuzzy controller we tested it in simulation, by using the ELDO program to verify the DC/DC converter behavior for different values of load, power and frequency. We thought to replace the error amplifier block of an existing DC/DC regulator with the developed fuzzy algorithm, and to simulate it both in continuous mode (with load of about 500 mA) and in discontinuous one (with load of 50 mA)

We have the following results:

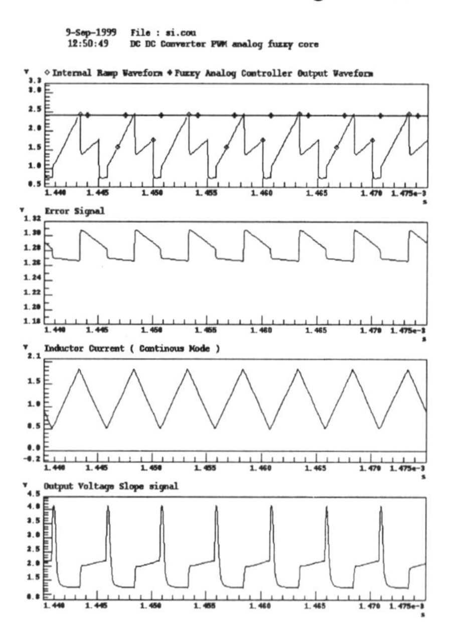

Figure 11: Results in CCM

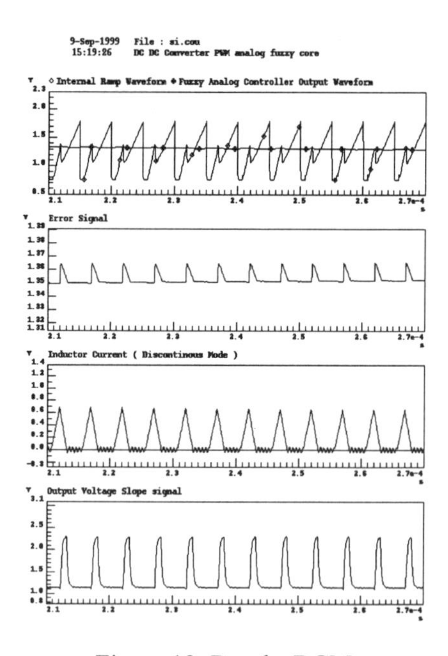

Figure 12: Results DCM

The implemented control scheme has the advantages of the *current mode* with ramp compensation and the advantages of a fuzzy control, being the error amplifier replaced by the fuzzy algorithm.

The fuzzy block confers robustness to the converter, allowing to realize a stable behavior also for the unacceptable values of a linear control. Opposite, the use of a fuzzy control determines the increase of the number of greatness whose variations will influence the behavior of the system: in fact the characteristic parameters of the fuzzy algorithm, as for instance those inherent the form and the position of the memberships, compete to determine the action of the control. The *first derivative* parameters, have been particularly object of reflection for the logical behavior of the DC/DC regulator.

The tests on the converter with the analogical implementation of the fuzzy controller have displayed a sufficient stable behavior, highlighting therefore the robustness of this type of non linear control. Maintaining stable the system, allows a mode variability of the inductance value. We remember in fact that elevated values of this parameter, move the operation point of the converter toward the *continuous mode*, bringing the system with a linear control type toward the instability.

Parameter variation in the fuzzy algorithm

The form and the values of the memberships individualize a number of parameters that define and characterize the fuzzy algorithm. Although their variations modify the control, the fact to operate and reference to a not well defined greatness (fuzzy) establish an implicit robustness of the algorithm to the parametric variations. Besides, we have worked with an analogical implementation, so, the number of parameters on which is possible and necessary to act to optimize the control, increase in relationship to the complexity of the same algorithm; this meaning that to the realization we have to consider for example the relation W/L of the MosFets, the precision of the signals used like references in the memberships generation or the numerous other variables that would not appear in a digital implementation. Actually, the research of the best values in the whole set of parameters has been conducted through a simulation study.

Conclusions

The present work shows that the use of an integrated fuzzy control in a current mode architecture allows to realize a step-up DC/DC converter characterized by high performances.

Although the fuzzy logic has been studied for control application in the converters, we suppose that how much developed in this work is important and innovative for solving converters problems.

In fact, the limits introduced by the traditional control are bypassed introducing in the PWM current mode scheme with ramp compensation, a fuzzy algorithm.

The developed fuzzy controller has been implemented in analogic; the aim is the capability to minimize the number and the dimensions of the components and allowing a realization on chip of an integrated DC/DC converter that includes in the architecture the control block, with least increase of area in comparison with the linear control circuits currently in use.

The obtained results, introducing the fuzzy algorithm, allow us optimistic applications in those problems in which it is necessary to increase robustness to the high variations of operative conditions in non-linear systems.

References

[1] D. M. Mitchell, *'DC/DC Switching Regulator Analysis'*, Mc Graw-Hill.
[2] B. Murari, F. Bertotti, G. A. Vignola, *'Smart Power ICs-Technologies and Applications'*, Springer 1996.
[3] A. Lionetto, *'Controllo Current-Mode in Logica Fuzzy per Regolatori a Commutazione in Configurazione Step Up'*, Tesi di Laurea, Dipartimento Elettrico, Elettronico e Sistemistico Università degli Studi di Catania.
[4] C. Burgio, P. Cammelli, P. G. Maranesi, V. Varoli, *'The DC Loop-Gain of PWM Voltage Regulators'*, IEEE PESC Rec., 1986.
[5] Linear Data Book, *'Modelling, Analysis and Compensation of the Current-Mode Converter'*, Unitrode Application note, 1995.

The Multispark System

Gaetana Mangiaratti

Accent S.r.l.

Via F. Gorgone 6, 95030 Catania - Italy

E-mail: gaetana.mangiaratti@accent.it

Introduction

The use of the electronics applied to the automotive is in constant growth and new solutions more and more innovative increase the content of intelligence allowing the attainment of performances, unthinkable some years ago, always much more competitive costs.

In particular the application of electronic control has completely revolutionised the traditional safety conceptions, comfort and performances in the automotive area.

Being the engine car the result of an assemblage of different components, the electronic applications could concern each component individual of the finished product, naturally according with the objective that we want to obtain, we choose an appropriate technology of semiconductors to optimise the performances, the cost and reliability.

Within the various applications, the control motor is particularly important and the electronic ignition is one of the applications that allowed the development of solutions that improve the performances and reduce the costs of the consumer (for example in the optimisation of the consumption).

To put in evidence this point we can mention the traditional system of mechanic ignition with contacts and compare it with the system of electronics ignition that is adopted in a lot of car engines.

To ignite the air/petrol mixture in the combustion chamber, it needs that between the electrodes of the spark plug a spark goes off.

It is desirable that such spark frees the greatest possible quantity of energy, to ignite all and well the mixture, so that the engine can distribute the maximum power [1].

The conventional ignition uses a contact breaker (platinized points) and a coil that is able to raise the voltage battery (12-13 V) up to a few kV, in order to set off the spark between the electrodes of the spark plug.

When the contact breaker is closed the coil stores energy through a flow of current from the battery, closes through the ground of the engine. When the contact is opening, the abrupt arrest of the current gives an extra-voltage that, applied to the primary coil and elevated from the coil ratio, allows to reach the necessary voltage to the breakdown of the dielectric between the electrodes of the spark plug where the spark goes off.

The principal problems derive from the usury of the platinezed points that, being subject to strong mechanical, electrical and thermal stresses, influence the system performances.

Moreover, it is not possible to increase the maximum energy stored because it would demand a value of current that would heat, corrode and blacken the contact surfaces, thus increasing the resistance and deteriorating both performances and reliability.

An other point to take into account is that, as the engine revolutions increase, the switching frequency of the points becomes very high and the coil current could not reach the necessary value to produce an optimal spark.

The reliability of the mechanical solution is seriously prejudged from the corrosion of the contacts, already evident after 3000-4000 km from their complete substitution, making difficult starting the engine at low temperature or when the engine is too hot.

In the course of any static proofs we could observe that with the engine in phase of starting or with idling engine, the voltage to the terminals of the spark plugs didn't reach the 8kV and this value was about at 5000-5500 rev/min. On the contrary, at around 2000-3000 rev/min this value becomes about 25 kV.

Listed previously the defects of a conventional ignition, we see how the electronic ignition resolves them.

The voltage impulse, furnished to the primary coil, is not more given from the interruption of the current that crosses the points. In the electronic ignition the contact drives the control terminal of an electronic device that, doesn't require a meaningful current and doesn't cause any usury of the platinazed points and then any phase displacement.

The voltage that is furnished to the primary coil is now controllable and it typically reaches values between 400V and 550V.

At this point there is a powerful spark able to ignite the combustion mixture completely and to maintain clean the electrodes of the spark plugs, the engine therefore will have a gain in performances, improving for example the acceleration phase from the low regimes, it will have less vibrations and acquires a better torque that allows to overcome the spurts in slope.

From the effected tests a voltage to the electrodes of the spark plug of around 45 KV is obtained independently of the regime of rotation of the engine.

In situations in which the engine's start could be difficult (for example in the winter or when the engine is too hot) one or two revolutions will be sufficient to start the motor also with batteries not perfectly loaded.

In the evolved systems the use of the platinazed points has been completely excluded also in the case of the management of the command of the electronic device.

In these more modern applications the synchronisation of the ignition system is demanded to a microprocessor that, through signals generated by special sensory places within the engine, manages completely all the phases from the storage of the optimal energy (that is function of the parameters of the engine, of the voltage of the battery and of the temperature) to the management of the operative frequency (function of the number of engine's revolution and of the environmental parameters).

After having shortly analysed these aspects, we could introduce what is the purpose of our work.

In fact, if the electronic ignition has allowed to overcome many of the limits of the traditional ignition, in a future more advanced systems that will allow to plan an electronic ignition device named "Multispark System" will be considered

This type of system must allow a rapid slope of the current of coil ignition (I_{COIL}) to facilitate the trigger of the spark; we foresee therefore a combustion of the mixture air/petrol better than traditional systems, in order to have better performances in terms of power. The current involved will be therefore fairly elevated.

We don't known now if such a device will be positioned in the coil or inside the central unit. Both the solutions present advantages and disadvantages. If it is put directly in the coil, we do not have any noise since neither the high current neither the troubles influence the central unit, but the management signal will not be directly available. On the contrary, in the second solution the signals are available but the noise would play a fundamental role.

The present work is therefore devoted to the realisation of a "Multispark" electronic ignition prototype capable to satisfy suitable specification.

The plan foresees that inside a certain interval of time more sparks for the ignition of the spark plug should be achieved, from which the denomination of Multispark.

A $V_{CONTROL}$ signal deriving from a microprocessor is applied at the input of the device. Such a system is made up by some circuital blocks such as the input stage, the driver, the voltage reference, the current limiter and the voltage flag.

Since specifications require a current of coil (I_{COIL}) of $15 \div 20$ A, we have chosen an IGBT power stage capable to support this current intensity. Moreover, application requires a very small coil in the order of $150 \div 200$ μH.

The Multispark System

In the modern electronic ignition based on inductive discharge, a μ-controller regulates the whole system, while a power transistor is used as a switch for allowing the primer of the spark.

The "Multispark" system takes into consideration most advanced applications of the automotive market in the field of the electronic ignition as the current of elevated intensity and a train of sparks instead of an individual one, to burn the mixture air/petrol well for the advantage of a better performance.

A possible scheme of this solution is shown in the Figure 1 where five external connections and a high voltage terminal connected to the primary of the ignition coil can be observed.

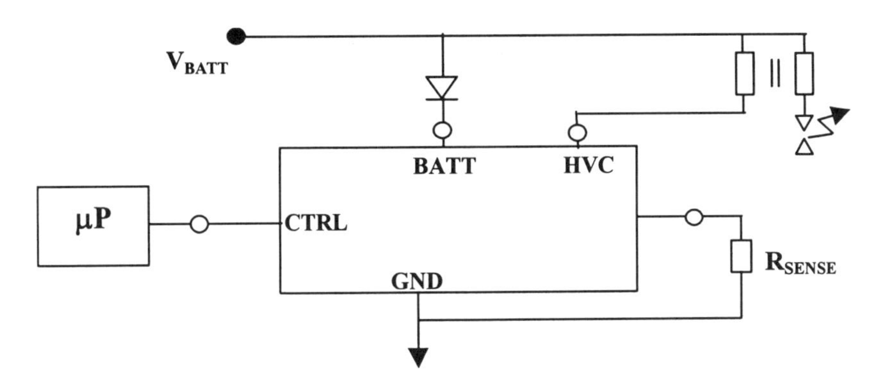

Figure 1: Scheme of external connections

The µ-controller sets the "ON phase" of the power stage by means of a control signal, $V_{CONTROL}$.

The SP signal, at the R_{SENSE} terminals, checks the value of the ignition coil current; V_{BATT} and GND are the supply signals deriving from the battery and the ground respectively, while the HVC is the high voltage terminal connected to primary of the ignition coil.

Block's diagram

In Figure 1 (see also Figure 3) the block's scheme is underlined. We observe that the device is supplied from a battery, whose voltage in each condition can vary from 6 to 24 V. An input signal $V_{CONTROL}$, managed from a µ-controller is applied to a comparator (input stage) that acts on the power stage through the Driver signal.

When $V_{CONTROL}$ reaches the V_{F1} threshold (reference voltage) the driver turns on and the current flows in the ignition coil. The Current Limiter compares a voltage reference with the voltage across a sense resistor, V_{SENSE}, in series with the coil. As soon as the current level reaches the desired value, the V_{SENSE} reaches the value of V_{F1} and the output of the comparator, the Current Limiter, turn on, switching off the Driver, through the transistor Q2.

Then the power device (IGBT) is switched off and the energy stored in the inductance of the primary ignition coil is transferred to the secondary producing a spark. During this phase the output of the Voltage Flag turns on maintaining the Driver switched off for all the duration of the spark, by the transistor Q1. The cycle restarts to completed discharge. The reference voltage and the current required by the device are generated from a Voltage Reference.

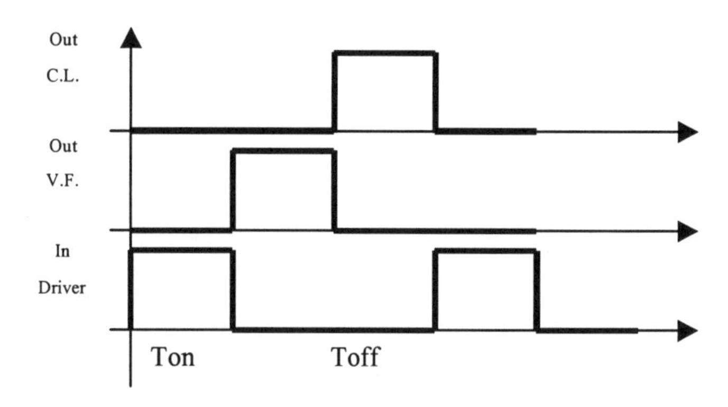

OUT C.L.	OUT V.F.	IN DRIVER
0	0	1
0	1	0
1	0	0

Figure 2: Sheme of intervals of time Ton - Toff

The strategy of the used control foresees that, while the $V_{CONTROL}$ signal is high, the Current Limiter regulates the Ton and than the Voltage flag fixes the Toff of the power device. A train of impulses of finite duration must be created to the output of the input stage allowing the turning on and the turning off of the Power respectively, and therefore the train of desired sparks.

The two transistors Q1 and Q2 realise a NOR function and the truth table is shown in Figure 2.

Naturally this is a simply scheme in which the intervals of time Ton and Toff don't reflect those real.

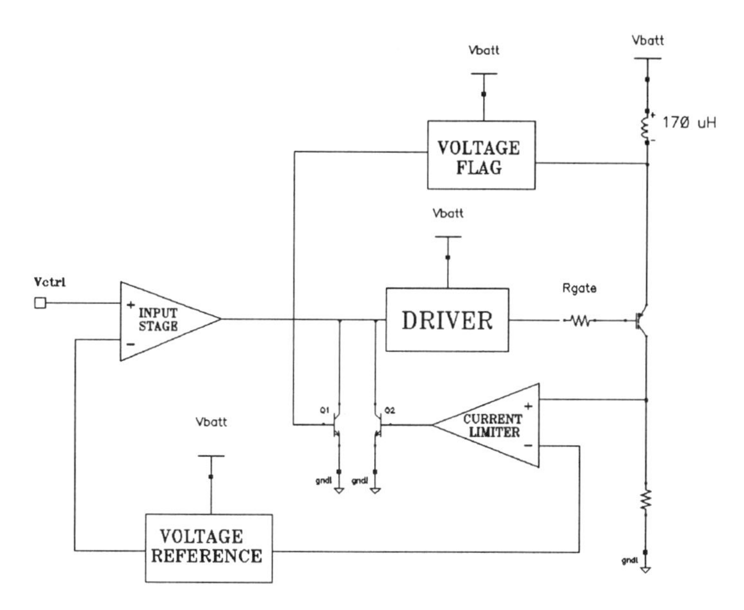

Figure 3: Block schematic

Voltage Reference

This circuit is based on a bandgap voltage reference with a VBG around 1.25 V that by means of an output stage is raised to the VREF value (R3V) equal to around 3V. Inside the circuit there are a signal BP2, which sets all the current sources of the Current Limiter, and a BP signal that sets current sources of the blocks Voltage Flag and Driver. Moreover, the signal R3V, equal to 3 V, is the voltage reference used to set all the threshold voltages of the input. It is a voltage kept constant as regards the variation of the battery voltage hypothesised in the plan (6÷24V).

The Voltage Reference generates a reference voltage V_{F1} that has the same law of variation in temperature of the resistance in series to the emitter of the power (IGBT). In fact VF1 is one of the inputs of the comparator for the limitation of the current. To the same comparator the resistance of sensing to the source of the IGBT is connected as shown in the Figure 4.

The R_{SENSE}, realised by a metal strip on the silicon chip, presents a temperature coefficient of 3900 ppm/°C, and depends on the temperature as follows:

$$V_{sense} = R_{sense}\, I_{coil} = R_{so}\,(1+\alpha\Delta T)\, I_{coil}$$

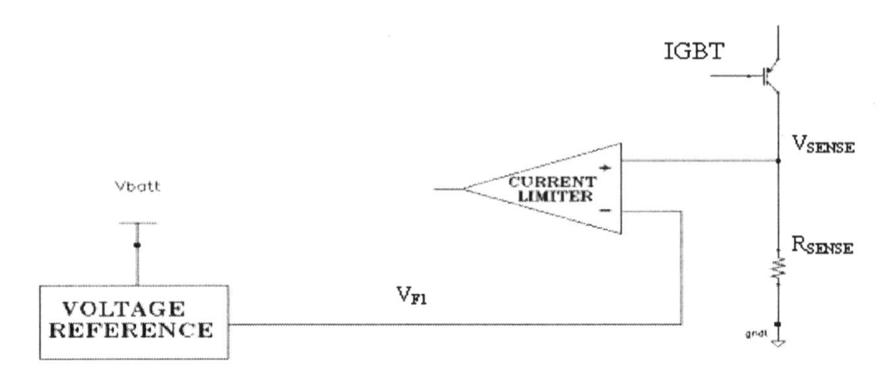

Figure 4: VF1 at the output of voltage refernce is compared with the drop voltage on the sensing resistance, V_{SENSE}

The problem becomes to guarantee that the voltage reference V_{F1} changes with the same law of V_{SENSE}, otherwise that it is able to follow the variations of the R_{SENSE}, compensating them. If we want activate the turning off of the power device to a constant value I_{COIL}, the temperature variation of V_{F1} will follow the variation of the R_{SENSE} and therefore of the V_{SENSE}. The circuit sizes setting $\frac{\partial V_{sense}}{\partial T} = \frac{\partial V_{F1}}{\partial T}$ where in the calculus of the expression has stayed approximately considered $\frac{\partial V_{BE}}{\partial T} = -2mV/°C$. Simulating the circuit with a coil of 170μH we have gotten an Icoil=21.8A @ 27°C. The value of the coil has been obtained by requirement, that they require a Ton=300μs.

Subsequently we made some simulations for different values of temperature, thus analysing the behaviour of I_{COIL} and the variation of V_{F1} in the maximum operative range for Automotive devices (40°C to +150°C). The simulation results are shown in Figure 5.

	-40 °C	**27 °C**	**150 °C**
V_{F1}	394 mV	524 mV	732 mV
I_{COIL}	22 A	21.8 A	20.5 A

Figure 5: Results of the simulations

The current I_{COIL} fits the values imposed by the specifications, where a variation of ±10% is accepted. As for as the variation of V_{F1} is concerned, it follows the voltage V_{SENSE} in the interval - 40°C and the 27°C. The more critical interval results to be between 27°C and 150°C.

Input Stage

To the input of the circuit, shown in Figure 1, we have applied a V_{IN} ($V_{CONTROL}$) signal deriving from a microprocessor, whose duration is typically of 2÷3ms. During this time interval the train of sparks is produced.

Inside there is a hysteresis comparator. In fact, the input of the comparator applied to a reference signal modifies its potential as soon as the V_{CTRL} reaches 2.9 V and it turn back to the previous state when the V_{CTRL} signal comes down to under 2 V. The hysteresis of about 900 mV avoids false commutations due to the noise.

Moreover, we have a signal deriving from the OR operation between the output of the two comparators (Current Limiter, Voltage Flag) and it let interdict the Power, acting on the Driver circuit, whose task is to switch the power device. In fact as soon as the output of the two comparators goes up (Figure 1) CL goes down, putting in interdiction Q10 and putting down the output signal, allowing the IGBT turn off. It is worth noting that in this contest $V_{CONTROL}$ is high.

In fact the innovation is to have more then one commutation for the power device, also having a high input signal.

Driver Stage

The Driver Stage, represented in the Figure 1 manages the interface between the input stage and the gate of the power device. At the instant of turning on of the Power the deriving signal from the Input Stage determines the turning off of the Darlington, allowing the gate signal going to the high level.

When I_{COIL}, increasing with the law defined by the inductive load, reaches the level of limitation, the signal at the input of the Driver goes low, since the Driver signal goes low too, allowing the interdiction of the Power. In the turning off circuit a Darlington has been used to get a fast commutation of the power device.

The Power stays off for all the duration of the spark because of a signal deriving from Voltage Flag, which maintain low the signal driver. At the end of the discharge the signal to the input of the Driver turns back high allowing the turning on of the Power.

This cycle is repeated for all the time in which $V_{CONTROL}$ stays high.

Current Limiter

The output of the current limiter acts on the Input Stage as soon as the current in the coil reaches the desired value, guaranteeing the turning off of the IGBT. The signal has gotten comparing the voltage drop on the sense resistor, that is situated between the source of the IGBT and the ground, with a voltage reference V_{F1} (as shown in Figure 2)

A comparator to the input of the device works through the unbalancing that is created on a current mirror. As soon as the threshold is reached (V_{SENSE}>V_{F1}), Q1 turns on and goes to disable the output of the Input Stage (see Figure 1). The Power turns off and the phase of spark is generated.

At this point we could think that the Power turns on again as soon as the output current, decreasing, carries the voltage on R_{SENSE} to an inferior value then the voltage reference V_{F1}. That doesn't happen since the control goes to the circuit of Voltage Flag that prevents the switching on again of the Power for all the duration of the discharge.

Voltage Flag

The Voltage Flag intervenes as soon as the Power goes off since in this phase the collector voltage of the IGBT rises due to the effect of the inductive load of the primary of the ignition coil. To assure a correct working of the circuit it is necessary that transistor Q1 turns on along with the spark happening (see Figure 1).

If for any reason the spark doesn't go off, the energy stored in the coil is dissipated on the Power because, for the protection of maximum voltage on the power device IGBT, as soon as the voltage of drain reaches around 450 V, the power switching on again.

A chain of zener put to the input of the circuit fixes the threshold of activation to around 50 V, these zener interface with the drain of the power device through the resistance of high voltage.

When the drain voltage of the Power reaches the 50 V (the spark has happened), thanks to the high current capability of some transistor, Q1 enters immediately in turning on, acting on the Input Stage output. This action maintains the Power off for all period of the extra-voltage.

Since this phase of extra -voltage is tied to the phenomenon of the spark, that produces oscillations on the voltage of drain and has an average value that depends on the voltage battery, it not possible to plan a fixed deactivation threshold, but on the contrary it has to be linked to the value of battery. The adopted strategy makes the current of the last stage in the circuit directly given by from the HVC.

Since after the first impulse on HVC the voltage of "sustaining" during the phase of spark is less than the threshold of 50 V, a latch has to be used to avoid the continuos output commutation.

This action is realised by a hysteresis comparator.

IGBT

The employment of the power device in most applications is needed when the device must behave as a switching. The switch transfers the energy stored in the capacitor to the primary of the ignition coil. This function is performed by a SCR. The switch is generally linked to a diode to conduct the reverse current. The ability of bearing high voltage in off state and high current in the on state, together with a high commutation speed represents the fundamental characteristics of such devices.

The majority of the electronic ignition adopts bipolar power devices, mainly in Darlington configuration. MOSFET can also be used but more complex production techniques are required.

A MOS device doesn't present the problem of bipolar transistors as for example the recombination of the minority carrier in the base, that limits the switching time [2].

The IGBT is able to give better performances, englobing the properties of the two above transistors, such as the driving voltage, the low R_{ON} for high voltage (BV> 500V) and the absence of secondary breakdown. Moreover, it can bear currents in the order of $15 \div 20$A and also more.

Since IGBT is a voltage driven component, it has obvious potential advantages when used in complex system. Its negligible drive energy allows the elimination of many components in the driver stage needed on the contrary in bipolar DGT discrete solutions. The low threshold feature, ensuring a full saturation at V_{GS} as low as 5V, gives an outstanding performance in terms of power dissipation (V_{DS}(sat)). Parts are available with and without internal voltage clamp [3].

In these applications the IGBT is a very attractive option in comparison to other types of power semiconductor, due to its high current handling capability and robustness, leading to smaller silicon and package size with consequent savings in board space and size.

Since the early 1970 when the bipolar Darlington transistors began to replace traditional mechanical switches, electronic switches have dominated automotive ignition.

The growth in the use of electronic equipment in the automobile coupled with the increasingly severe emission regulations, have required a more sophisticated approach from the electronic ignition dcsign point of view and greater demands have been placed on the switch.

Distributorless ignition system use either a dedicated coil for each cylinder, or one coil for a pair of cylinders. It is difficult to use the bipolar Darlington as switch in this application, due to the complexity of the required driving circuit. The potential advantages of the use of voltage driven components, which require negligible drive energy, are evident.

Additionally, the introduction of low threshold devices eliminates one important drawback of the use of IGBT, when the engine is cranked the battery voltage can fall down to 6 V or less, which would not be sufficient to fully saturate standard IGBTS. Logic Level device with a threshold below 2V ensure a full saturation even with a drive voltage well below 5 V.

The distributorless ignition system that we consider in the design is the ignition system with inductive discharge as shown in Figure 6.

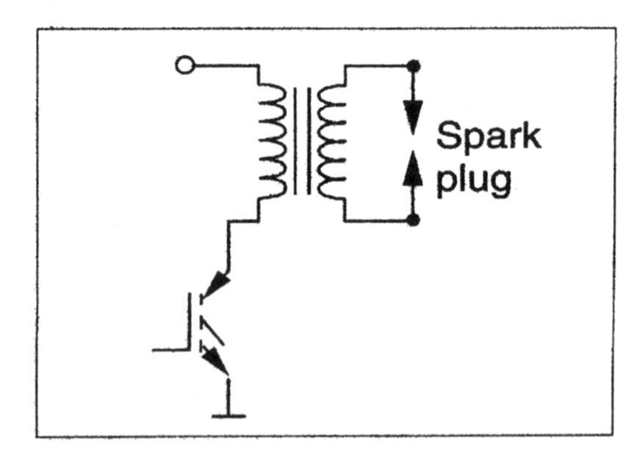

Figure 6: Ignition system with inductive discarge

At the turn-off, the collector voltage of the IGBT can easily exceed its breakdown voltage, for example due to the delay in the ionisation of the spark-plug, or accidental disconnection of the spark-plug.

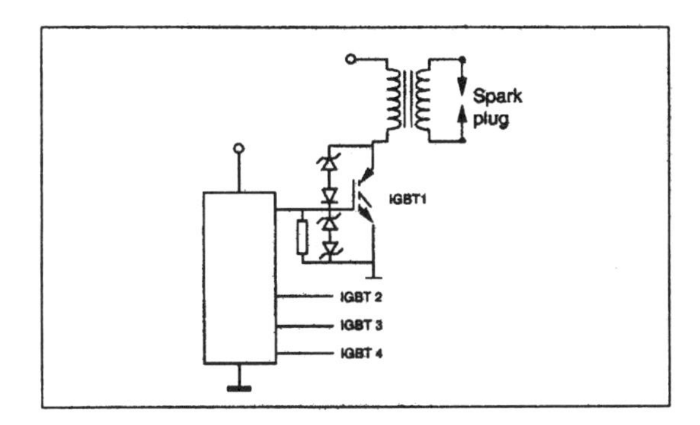

Figure 7: System with clamping circuit into the IGBT

Consequently to avoid the destruction of the device a clamping circuit must be used. Varied ways to clamping the device exist. The clamp could be external, integrated in the circuit of pilotage or integrated into the IGBT.

The IGBT used in the design is the LZB5, it is a device produced by ST.

The circuit of clamp is inside, and it has constituted by a zener of 400V, positioned between the gate and the drain. The high voltage zener integrated into the IGBT eliminates the need for an external clamping circuit (see Figure 7).

A solution of this type requires a more expensive switch, in fact one more mask level is required to integrate the clamping zener diode with the IGBT. However this method increases the reliability of the system.

The current and voltage behaviour of an IGBT in a inductive discharge ignition are shown below in Figure 8:

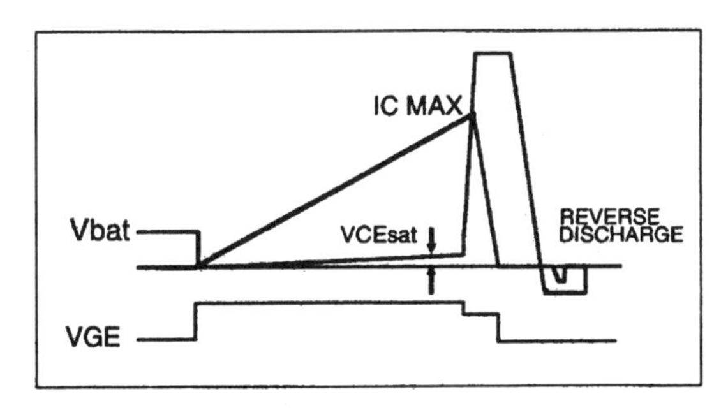

Figure 8: Behaviour of current and voltage in a inductive discharge ignition

Simulations

The simulations concern with the general circuit. The following waveforms are taken into account:
input voltage ($V_{CONTROL}$) deriving from the micro-controller
input voltage of the Driver (V_{DRIVER});
current in the coil (I_{COIL});
voltage on the high tension collector ($V_{COLLECTOR}$).

From the graphics, reported in Figure 9, it is evident that during the interval time of 2ms, generated by the microcontroller, the input voltage of the Driver is a signal of period T=Ton+Toff and the train of desired sparks is gotten.

This is evidenced in the graph that shows the current in the coil (I_{COIL}). It is possible to notice that the current reaches a value of 21.8 A. The voltage on the collector is clamped at about 400 V.

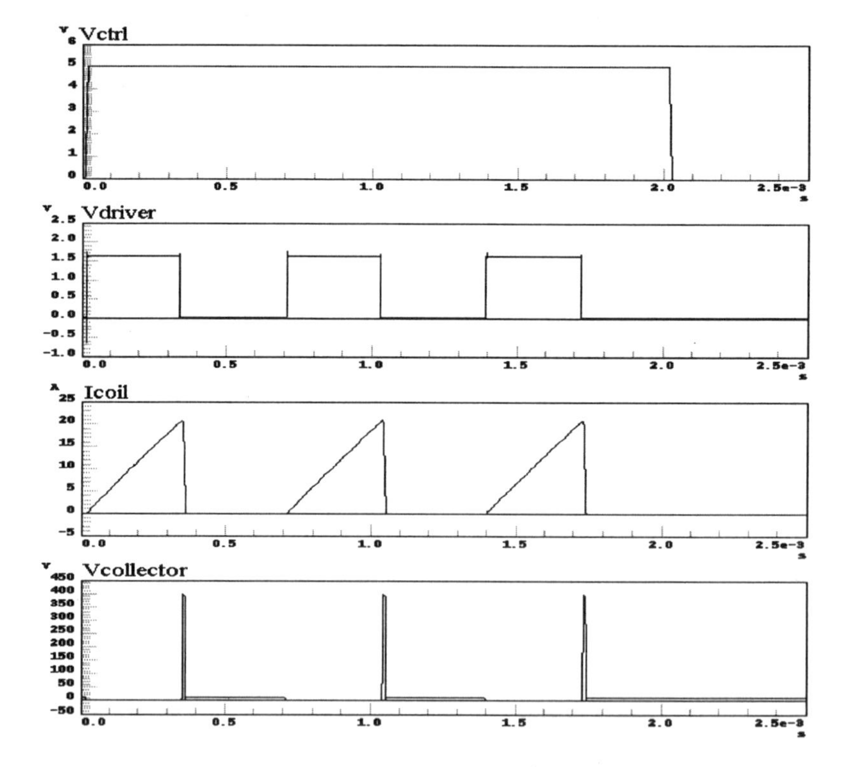

Figure 9: Simulation respective to V_{Ctrl} V_{Driver} I_{Coil} and $V_{Collector}$

Experimental board

To experimentally estimate the circuit performances, a board has been implemented by using devices already existing in ST like the VB130 and the Smart IGBT.
Specifically only the Voltage Flag block of the VB130 and the Driver Stage of the Smart IGBT were used, while the rest of the circuitry was properly adopted.

It is important to notice that the IGBT and the R_{SENSE} shown in Figure 10 are really inside to the Smart IGBT. In fact, we find two pins inside the device, Positive Sensing and SP-igbt, the first represents the output of the Driver, while the second is the terminal to potential high of the sensing resistance.

A PWL signal of 5V amplitude and duration of 2ms was applied at the input of the device.

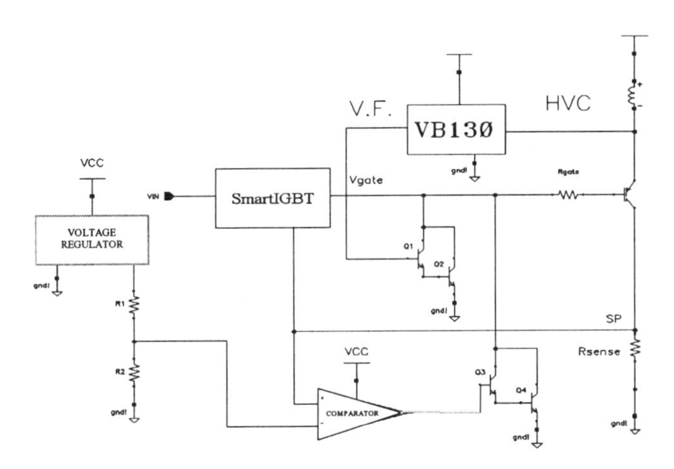

Figure 10: Board assembled in laboratory to simulate the Multispark

In Figure 5, we can notice the wave forms related to the input voltage, to the current in the coil and to the IGBT drain voltage.

According to the graphic of Figure 3, we can observe that inside to the input signal a series of sparks is gotten. You note that simulator waveform are in a perfect agreement with the experimental waveform.

The value of current is less than 21.8A since we used a 1mH ignition coil, bigger than that used in simulations (170uH) like it has been required by the specifications.

It is worth noting that the current waveform never reaches the zro value, since we have simulated the circuit with the secondary of the ignition coil floating, without the spark plug. For such reason in the graph you cannot see the Toff, since it depends on the discharge period.

Sorts these premises, therefore it is possible conclude that the proofs in laboratory have confirmed the results gotten in simulation.

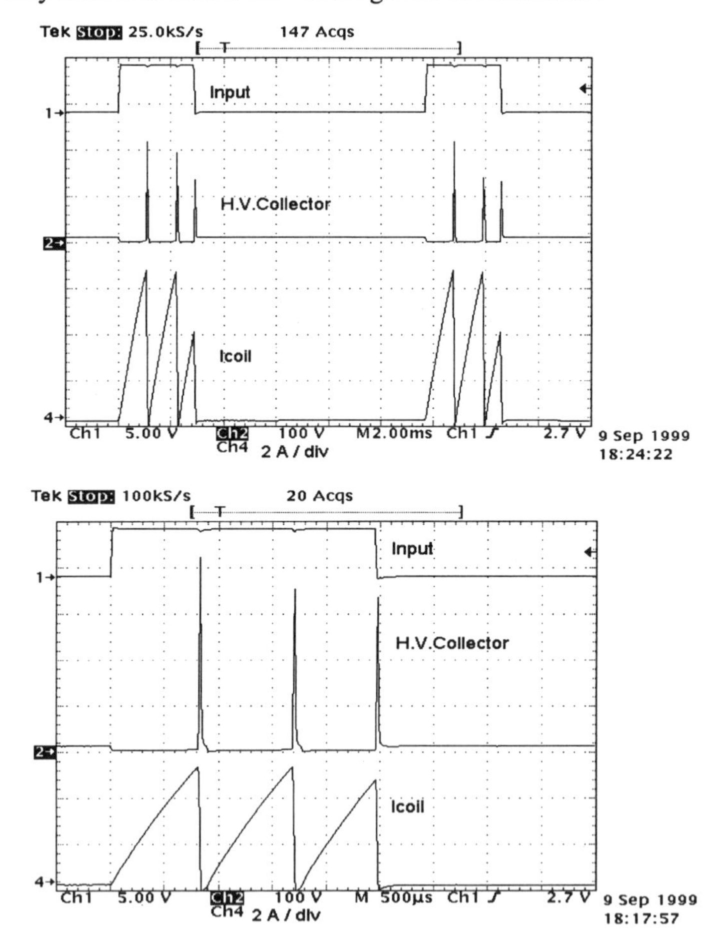

Figure 11: Experimantal results

Conclusions

As it is shown in Figure 11, where the current trends of the ignition system are reported, the main tasks for electronic ignition of a Multispark scheme have been reached.

The general circuit must be completed with the addition of an other block, the Load Dump Protection, that is useful for protecting the device from the "load dump" due to the fact that the circuit is directly supplied from a battery that operates in a range of values from 6 to 24 V.

The integration of the system into a single hybrid chip circuit is expected.

References

[1] Ned Mohan, Tore M. Undeland, William P. Robbins: *"Power Electronics: Converters, Applications, and Design"*. John Wiley & Sons, Inc., New York 1989.

[2] K. Kit Sum: *"Switch-mode Power Conversion: Basic Theory and Design"*, Marcel Dekker, Inc., New York, 1984.

[3] K. Thorborg: *"Power Electronics"*, Prentice Hall International (UK) Lmt, 1998.